Peter Calvert worked professionally as an engineering research associate. He has extensive experience in Buddhist Vipassana meditation, Spiritualist platform mediumship and holotropic breathwork. After learning to write while in a meditative state, he began communicating with non-embodied identities who sought to transfer their knowledge into written form. The year 1998 proved a watershed. Peter began leading meditation groups, and also started channelling material intended to provide a new perspective on spiritual practices and metaphysical ideas suited to contemporary seekers. This perspective included new terminology, metaphors and models that update ancient spiritual understanding as described in the Hindu, Buddhist and Taoist traditions. Peter has established Agapeschoolinz to share what he is discovering, and is the author of a number of books on spiritual topics, including *The Matapaua Conversations* and *The Kosmic Web* with Keith Hill.

Keith Hill is a writer whose work explores the boundaries between spirituality, science, history, religion, culture and psychology. His spiritual training was in the Fourth Way teaching of G.I. Gurdjieff. More recently he has been drawn to the Michael Teachings. His books include *Striving To Be Human*, *The God Revolution* and *Practical Spirituality*, each of which won the Ashton Wylie Award, New Zealand's premiere prize for spiritual writing. Keith's translations include the *Bhagavad Gita* and *Interpretations of Desire*, mystical love poems by the Sufi master, Ibn 'Arabi. Keith's latest non-fiction work is *The New Mysticism*.

A NON-EMBODIED SPIRITUAL PERSPECTIVE
CHANNELLED BY PETER CALVERT & KEITH HILL

PETER CALVERT & KEITH HILL
The Matapaua Conversations
The Kosmic Web

PETER CALVERT
Guided Healing
Agapé and the Hierarchy of Love

KEITH HILL
Experimental Spirituality
Practical Spirituality
Psychological Spirituality
What Is Really Going On?
Where Do I Go When I Meditate?
How Did I End Up Here?

RELATED BOOKS

PETER CALVERT, RICHARD BENTLEY
CAROLYN LONGDEN, TRISHA WREN
People of the Earth

KEITH HILL
The New Mysticism

LEARNING WHO YOU ARE

AN INTRODUCTION TO EXPERIMENTAL SPIRITUALITY

PETER CALVERT & KEITH HILL

First edition published in 2019 by Attar Books
Auckland, New Zealand.

Copyright © Peter Calvert & Keith Hill 2019

The right of Peter Calvert and Keith Hill to be identified as the joint authors of this work is asserted according to Section 96 of the Copyright Act 1996.

Paperback ISBN 978-0-9951204-4-0
Ebook ISBN 978-0-9951204-7-1

All rights reserved. Copying and distributing passages excerpted from this book for the purpose of sharing and discussing is permitted on the condition that the source of each excerpt is fully acknowledged, and such excerpts are not onsold. Otherwise, except for fair dealing or brief passages quoted in a newspaper, magazine, radio, television or internet review, no part of this book may be reproduced in any form or by any means, or in any form of binding or cover other than that in which it is published, without permission in writing from the Publisher. This same condition is imposed on any subsequent purchaser.

Cover image by Frankie/Shutterstock

AgapeSchoolinz
www.agapeschool.nz

Attar Books is a New Zealand publisher which focuses on work that explores today's spiritual experiences, culture, concepts and practices. For more information visit our website:

Attar Books
www.attarbooks.com

CONTENTS

	Preface	7
	Introduction	11
1	Dao and the Multiverse	17
2	Nodes Of Dao-Consciousness	25
3	Each Life is an Experiment	33
4	Accumulating Human Identity	40
5	Learning Who You Are	48
6	The Process of Self-Enquiry	55
7	The Function of the Aura	62
8	The Art of Meditation	70
9	Agapéic Space and its Correlates	76
10	The Agapé Frequency Scale	84
11	Life in the Multiverse	90
	Afterword	98
	Further Reading	99
	Index	119

PREFACE

PETER CALVERT AND I FIRST MET in 2008, at a book fair in Tauranga, New Zealand, where Peter was presenting his first channelled work, *Agapé and the Hierarchy of Love*. I bought a copy, started reading, and was immediately drawn in. The material offered a rational, scientifically-oriented approach to spiritual development that made a lot of sense. But as much as the text, I was drawn to the standpoint from which it was written. The identities Peter was channelling were looking down on human existence from *way* above. They were knowledgeable without being dogmatic, exact without being stuffy, and they had a sense of humour. What especially attracted me was that they offered a relevant contemporary take on metaphysical and spiritual matters.

As a result, I offered to help Peter edit his next channelled book, *Guided Healing*. While working on it I decided meeting Peter presented a unique opportunity to ask a group of non-embodied identities, if that's who he was communicating with, some serious questions. I consequently put together a list of what became one hundred questions on "big" topics, such as the creation of the universe, how the spiritual and physical connect, and the purpose of our life on Earth. The answers Peter received surprised us both. They went far beyond what either of us had anticipated, and opened up an entirely new—at least, for us—way of conceiving reality.

The answers Peter received, and the diary he kept while working on them, I edited into our first collaborative book, *The Matapaua*

Conversations. Since then our collaboration has blossomed, resulting in eleven books to date, channelled by each of us. Beyond this, during the past three decades Peter has channelled well over a million words (and counting), which he has archived on the www.agapeschool.nz website. I have produced a much more modest volume of material.

Learning Who You Are is an introduction to these channelled texts. It is designed to offer an easy-to-read overview of what the guides, as we denote them, have shared with us to date. In these pages you will find material that deals with the topics traditionally explored in spiritual teachings, including metaphysics, human spiritual nature, psychospiritual development, meditation, and the aura.

What differentiates this material from traditional teachings is that its conceptual frame is non-religious, preferring to link spiritual processes to current scientific and cultural knowledge. It explores the spiritual aspects of our existence in ways that sync with how we see the world in the twenty-first century.

All the material that follows has been channelled by Peter or by me. It is drawn from a range of sources. Some has previously been published in our books and online, some was channelled for group meetings and appears in print for the first time, and a little interlinking material has been channelled specifically for this publication. An appendix lists further reading that expands on the material presented here.

Keith Hill, editor
Auckland, July 2019

LEARNING WHO YOU ARE

INTRODUCTION

THIS BOOK IS INTENDED TO CLARIFY the relationship between the physical and the spiritual, along with the nature of human identity that participates simultaneously in both. We will discuss this relationship using concepts, metaphors, models and terminology appropriate to twenty-first century thought. Much of what we are offering in these pages has a scientific flavour. Our purpose in adopting a scientific approach is to generate material that maps onto what is generally agreed regarding the nature of reality in the early twenty-first century. In this way we hope that what we are offering will be useful to contemporary explorers enquiring into what has traditionally been known as spiritual knowledge.

For millennia, discussion of spiritual matters has been distorted by the idea that the spiritual is special. In fact, neither the spiritual domain nor spiritual identity are special. Neither is elevated. To speak somewhat gnomically, they are what they are, just as the physical domain and the bodies that live in it are what they are.

Despite this, today people continue to be told, especially those in religious circles, that the spiritual is special. However, contradicting this assertion, most people feel ordinary within themselves. Because they feel ordinary, yet keep hearing discussions concerning the elevated nature of spiritual identity and spiritual attributes, the likelihood increases that they will feel uncomfortable discussing spiritual matters. Perhaps they will even feel unworthy of participating in such discussions. As a result, they are unable to appreciate that their essential nature is spiritual. This is an error, promulgated generation after generation since time immemorial. Our intention in what follows is not just to correct such an erroneous view, but to offer a fresh way of conceiving of spiritual nature that integrates it with everyday existence.

The relationship between spiritual existence and physical existence is actually quite straightforward. Imagine standing on a river bank. You cross the river to the far bank, live there for a time, then cross back to the original bank. This illustrates an individual's transition into the state of embodiment, of being a spiritual identity living for a time in a body, then passing back out of it. No change of essential nature occurs. All that is involved is a change of location. People remain who they are wherever they are.

By describing the relationship between the spiritual and the physical in this way, we wish to make clear that our approach centres on human experience. We are emphasising an empirical approach to spiritual matters, offering models and explanations that may be verified via personal experience.

This raises an important point. All models and explanations are by their nature partial, because reality is multi-layered and complex and models and explanations necessarily present simplified views of reality. Being partial and simplified, they are discarded when no longer useful. This is standard procedure in the sciences. The scientific method requires researchers to gather data via empirical observations, then propose theories and develop models that explain as much of the data as is possible. Then, as further observations are made, and more data is gathered, the current theories and models are confirmed, or they are adjusted to accommodate the new data, or they are discarded because they no longer incorporate all that is being discovered. New models and theories are then developed that more fully incorporate all the data.

We expect the same process to apply to our explanations and models. They are designed to fit with the data gathered by current researchers. And they are offered with two goals. The first is to expand on what is already known. We do so by offering new ways to view what is currently understood, particularly by making connections between apparently disparate aspects of human experience and knowledge, showing that they relate in ways that have previously been only partially understood, or not understood at all. Second, our intention is to suggest new directions for spiritual enquirers to explore, which we expect will prove illuminating and useful to them. Yet behind all we propose is the expectation that in due course it will be absorbed into new levels of future understanding, or discarded as future enquirers obtain deeper levels of insight and require fresh models and explanations that incorporate their new perceptions and insights. This is how scientific enquiry progresses. This is also how we propose spiritual enquiry may most usefully progress today.

We acknowledge that the universal adoption of this process for spiritual enquiry is a long-term goal. Currently, ancient beliefs continue to have a powerful hold on people's concept of the spiritual. In particular, the idea that the spiritual is special by virtue of being divine limits people's ability to adopt the open-ended, empirically-based, exploratory approach to spiritual matters we are proposing.

Rejection of the idea that the spiritual is divine will not be easily achieved because it is so deeply embedded in all the world's religions. It has been historically and cross-culturally endorsed for millennia, making it a universally shared concept. To use a commonplace saying, that makes it a tough nut to crack. Yet the idea that the spiritual is divine is responsible for the radical segregation of the spiritual from everyday human existence. From the perspective of the embodied human being, the spiritual is nebulous, mysterious and difficult to comprehend. Even among those who accept the spiritual exists, it is usually viewed as cut off from everyday experience. Yet from the perspective of the spiritual realm, a continuity extends between the spiritual and physical domains that is entirely natural. From that perspective, segregating the spiritual from the physical is absurd.

The task we have taken on is to demythologise, naturalise and normalise the relationship between the physical and spiritual domains. By discarding the notion of the divine, we—by which we mean, you in a physically embodied state, and us in a non-embodied, spiritual state—

have the opportunity to begin a fresh conversation regarding human identity and the wider reality in which we all dwell. That we can have such a conversation free of the usual bric-a-brac of notions accumulated over the millennia is rare. That we can have this conversation in a cultural context in which participants have access to vast stores of relevant knowledge, and will not be persecuted for pursuing lines of enquiry that wander far from the norm, is rarer still.

This being the case, and even though we well know that realistically these books will have little immediate impact, we consider this too good an opportunity to pass up. In stating this, we acknowledge our fellow contributors, Peter Calvert and Keith Hill, have reached the same conclusion and have joined us in exploiting this period in human history to explore new ways of describing fundamental spiritual experiences and concepts. Without their help the expansive channelled materials, and the growing number of books derived from them, would not exist.

Given that our teaching regarding the connection of the physical and the spiritual has to date been largely expressed in a number of quite technical books, we have come to the view that what is needed is a general introductory book that presents, in a non-technical, accessible way, an overview that weaves together the diverse strands of what has been channelled to date. The diversity has arisen because when we work with embodied individuals we necessarily operate in the context of the skills they have developed, the knowledge they have accrued, the openness (and blindspots) of their outlook, and the goals they have selected for this life. Each of our two collaborators has his own particular approach, which is reflected in the kinds of experiences he is interested in exploring and the kinds of materials he is best positioned to process.

Peter Calvert is a meditator who is skilled in a range of mediumistic activities. His formal training is in the sciences and engineering. Combined with his aversion to religion, along with much else too extensive to list here, this has resulted in an individual who is very much in sync with our goal of offering a demythologised, naturalised approach to spirituality. Keith Hill's experience has a more psychospiritual leaning. University educated, like Peter, his spiritual training emphasised the need to grapple with psychological traits and develop all aspects of the psyche. Meditation and psychological transformation are complementary aspects of spiritual growth. Each is more effective when practised in tandem with the other. So their partnership is valuable on multiple levels, for them personally, and for the accomplishment of our joint task.

That two people are working together to generate this material is unusual. Most channels work alone. The advantage in this case is that our collaborators are able to offer each other support. Not that it is much needed, but not working alone has a psychological impact. Each also has individual strengths, which are complementary. It means that the channelled material for which they are the vehicles can be more complex and varied than is usually the case. This is a significant advantage to us, because the different kinds of materials we wish to communicate can be directed towards one or the other as appropriate.

We are similarly diverse. Each of our collaborators is working with a separate non-embodied entity. The two of us are reintegrated beings. We are reintegrated in the sense that we each comprise hundreds of separate human identities who have completed all their incarnational cycles. As reintegrated identities we are now working together to fulfil a collective task. However, besides the two of us who are working with Peter Calvert and Keith Hill, other reintegrated, non-embodied entities similar to us are simultaneously involved in other aspects of what is a long-term, multi-faceted project.

So much organisation is required because what is being initiated is an intervention in human culture, whereby demythologised spiritual concepts are being offered via a large number of individuals, across many countries, and over an extended period of time. So this series of channelled materials is part of an extensive range of communications and activities involving many people over many human centuries.

We note that neither of these two individuals will necessarily be involved in this same task in their next lives, given they each have their own issues to work through, karma to address, and talents to progress. An opportunity to take part in this undertaking arose and they were available and appropriate collaborators. They have skills and intents that mesh with our expertise and intents. We state this to make clear that these individuals are not special. Equally neither are we divine. Nor are you who is reading this. We are all what we are, doing what we do, working through what currently draws our attention and requires our effort.

We conclude with a comment on our calling the approach advocated here experimental spirituality. We asked our scribe to look up the definition of experiment: "A scientific procedure undertaken to make a discovery, test a hypothesis, or demonstrate a known fact." This definition applies to the process of spiritual exploration as we discuss it in these pages. The purpose of spiritual practices are to lead you to make new discoveries within yourself and in the world around you. Elsewhere we have suggested that beliefs be treated not as truths but as propositions, and that they be tested in the laboratory of personal experience. That equally applies to what we state here. Our statements are propositions that you, our readers, can only make real for yourself by testing them in the context of your life. The explanations and models on offer can only be seen as valid if they illuminate your current life experience or open up new avenues for exploration. Then you are in a position to empirically demonstrate to yourself that they reflect your experience and so are worth sharing with others.

To date the books we have produced with the assistance of our two collaborators have dealt with three principal topics: metaphysical overviews, meditation and related issues, and psychospiritual self-transformation. All three topics have been explored within a developmental context, that is, in relation to how understanding them assists spiritual enquirers' growth. In this book we combine these topics to create an overview that introduces the full scope of what we are offering.

We wish you, our readers, a fruitful interaction with these ideas. They are offered in the hope that you may extract from them something that is illuminating and useful, and that they stimulate your own growth and explorations. For those who do find them illuminating, more books are available that enter at some depth into the topics introduced here.

CHAPTER 1

DAO AND THE UNIVERSE

W E BEGIN OUR EXPOSITION WITH WHAT HUMANITY identifies as the beginning, the origin of the universe. That origin is well known to human scientists in a purely physical sense. The emergence of a singularity and the big bang that followed is an adequate explanation for what happened to bring this universe into existence. What is not well known is the spiritual intent behind that coming to be.

What we will attempt to describe here is something of that intent and how it has created the universe—for creation there certainly was. However, creation did not occur in the way that is traditionally and religiously understood. God was not involved in the sense that is proposed in the world's religious literature. In order to explain the process of creation, we replace the traditional concepts of God, which has become over-used and outworn, with the term of the Dao.

The Dao is the ultimate unmanifest. Everything that exists comes from the Dao. Consciousness is intrinsic to the Dao and derives from it, as does intent. The Dao in its intent is magnanimous and expansive in a way that those involved in human existence, focused on the much narrower concerns of the physical self, are incapable of manifesting.

Inextricably bound into the Dao's intent is an active force to which we give the name *observer*. In simple terms, it could be said that the Dao is the ultimate source of all that is and the *observer* is its intent. The *observer* is filled with goodwill, intelligence and love, plus many other qualities that cannot be expressed in human language, because there is no equivalent experience in the human domain. It must be understood that the *observer* is not a personal manifestation of any kind. When thinking of the *observer*, think of an abstract intent that cannot be fully described. It is an urge. A momentum. That is all. Certainly, do not conjure in your mind the image of a being of any kind. It is not God in a traditional religious sense.

We have chosen the words Dao and *observer* because they are relatively neutral terms, at least in the context of secularised Western English language. In doing so, we are attempting to avoid religious God language. So while aspects of what we are saying here have certainly been described before, in other cultures and eras, generally in religious literature, we are developing and updating those older concepts, constructing culturally relevant models and using terminology appropriate to these times.

It was via the *observer* that the Dao created the physical universe. When discussing this it is necessary to expand the discussion to include a cluster of universes. To appreciate what is involved, picture an extensive field within which multiple universes are expanding and contracting. Science calls this the multiverse model. We affirm that this model is valid, although not in the way cosmologists currently speculate.

How many universes exist within the multiverse? We affirm between five and ten universes are currently going through their cycles of expansion and contraction. We cannot assign a particular number because the human timescale is simply too truncated to accommodate the vastness of what is occurring. Humanity lives in a *now* that simply does not exist for us. Or for the multiverse. Or the *observer*. We exist outside the spacetime continuum humanity lives within. For us to generate a sample, which is what we would need to do, by making a theoretical slice through spacetime and then use that as a reference moment to definitively state, "As of *now* x number of universes exist within the multiverse," is ultimately an arbitrary exercise. There is no *now*. There is a range of current existence. From which we can derive an approximation only.

Each universe has specific parameters within which it exists. Just as scientists have identified certain numerical values in relation to this universe, such as the strength of gravity and the values of the weak and strong forces, and they know that these values are basic to this universe being what it is, so the parameter values basic to other universes are different, and consequently give rise to a different universal nature and conditions for existence.

Accordingly, it can be seen that the universe humanity exists within is actually part of a multiple initiative. It is not a sole attempt, as the philosophers say, to conjure something from nothing. It is part of a wider experiment to explore many varied possibilities of existence. And we say *experiment* advisedly, because this is certainly one way the multiverse may be viewed.

The *observer* initiated the multiverse experiment. The *observer* selected the experimental parameters for each universe, then set them in motion. Given each universe is at a different phase within its complete repeated cycles of expansions and contractions, some will cease to be before others. They will, or will not, be replaced by other universes according to the *observer's* intent. So this initiative identified as the multiverse is an ongoing process, the scale of which is beyond the understanding of any physical mind. It is not a randomised initiative. An infinity of universes has not been brought into existence out of a vague hope that one universe out of millions might succeed. It is an intelligently intended and constructed experiment.

Many scientists deny that there is any purpose to the universe and what exists in it, that everything happens accidentally, entirely as a matter of chance. They adopt this view because the only coherent alternative view is the traditional religious view, which is that God created everything and now personally oversees and controls it. Neither religious metaphysics nor scientific naturalism are accurate. As a corrective to partial thinking, we present the notion that the intent of the *observer* certainly led to the multiverse coming into existence in what could be said to be a creative act. But no overseeing or control is subsequently exerted in the traditional religious sense.

The *observer* has never been involved in any physical universe. Nor will it ever be. After setting the underlying parameter values for each universe, the *observer* has stood back, to use another inadequate human metaphor, and observed what occurs. This is somewhat like the Deist notion that a personal God created the world then stood back and let creation function according to natural laws. Yet the analogy works only partially. We repeat, the *observer* is nothing like a personal God or a person of any kind.

To come back to the origin of this universe, and to use current scientific terminology, following the big bang the laws of physics spontaneously transformed energy into matter, matter developed into gas clouds, gravitational forces drew the gas into stars and planets, and on certain of these planets environmental niches formed, some of which were supportive of life, from which life eventually emerged.

Behind the creation of each universe is a multiplexed intent to generate environments in which ever more varied and complex forms of physical life come into existence and evolve. But throughout this extensive process there is never any intent, at the *observer's* level, to control either the process or the outcome. From the *observer's* perspective there is also no attachment to what happens to any single life form.

So the multiverse is an extended experiment of inconceivable magnitude in which sets of opportunities have been constructed for the purpose of facilitating the emergence of ever more varied life forms. There is benign goodwill at the outcome of each universe. But no outcome is predetermined.

The creative act of something coming to be out of nothing, or at least out of no other material substance, has long puzzled humanity's best minds. In order to understand what creation is, it is necessary to extend one's conceptualising beyond the electrophysical and embrace the notion of the electrospiritual.

When the Dao gathered an intention within itself to manifest the multiverse, an intention we have characterised as the *observer*, what happened can be likened to a fluctuation. That fluctuation manifested the electrospiritual. Within the electrospiritual were encoded patterns of information. That information spontaneously formed a singularity that resulted in the coming into existence of this universe.

In all, we identify three layers or levels: the electrospiritual, the electromagnetic and the electrophysical. The electrophysical refers to physical bodies constituted of chemicals and electrical impulses within the nervous system. The electromagnetic, constituted of microwaves, x-rays and the like, is well known. The electrospiritual is postulated as an extension of these frequency fields. Just as the electromagnetic consists of finer frequencies than the electrophysical, so the electrospiritual consists of finer frequencies than the electromagnetic. Patterning across all three connects the manifest physical multiverse and the creatures and structures within it with the initiating creative and wholly spiritual intent of the Dao.

The physicist David Bohm has named a subtle component of reality the implicate order. We here adopt his term and use it to refer to the patterns of information embedded within the electrospiritual. The electrospiritual provides patterning to each and every living being in the electrophysical domain. The implicate order within the electrospiritual contains specific blueprints, according to which the cells that constitute the bodily makeup of each species conform.

Every living creature's body, whether it be an amoeba, a fungus, a plant, an insect, a crustacean, or an animal, is a configuration of the electrophysical into a distinctive biological form. Each species' distinctive bodily form, and hence the bodily form of every individual within each species, is a result of patterning that comes from the implicate order. The human body, to use it as an example, begins its life as a zygote formed by an egg and the sperm that fertilises it. The zygote's cells rapidly multiply and the developing foetus grows. Some cells become flesh, others nerves, some form the backbone, yet others the kidneys, heart, lungs and the brain. This process involves undifferentiated cells turning into specialised cells that graduate to specific places on the endoskeleton and perform particular functions within the growing body. DNA conveys traits to the developing body at the genetic level. But we maintain that the overall body is structured not entirely by DNA but that significant information is conveyed from the electrospiritual implicate order that tells the growing foetus' cells where to go and what to become. In effect, the cells conform to a higher level ordering dictated at the implicate level.

Clearly, this is a contentious statement. It will take considerable research to confirm. Eventually it will be done. At this stage what we have asserted can only exist as an unconfirmed proposition—alongside much else proposed here.

In summary: The electrospiritual provides patterning to each and every living being in the electrophysical domain. The implicate order within the electrospiritual contains specific blueprints, according to which the cells that constitute the bodily makeup of each species conform. Within these blueprints is the intent of the *observer*. It is too simplistic to say that the *observer* designed each blueprint in the implicate order. It did not, and we are not saying so. A much more layered, subtle and creative process is involved in ordering at the implicate level, which involves the intent of nodes of Dao-consciousness. *[Editor: For nodes see the next chapter.]*

This must be understood to be an interpenetrated model, in which the *observer*, the connecting patterning, and the universes interpenetrate each other. Ken Wilber's notion of holons is appropriate here, with reality consisting of a number of nested layers that are simultaneously individual parts and inextricably parts of the whole.

What we are trying to describe cannot be described. The danger is that the human mind automatically reduces any concept to human size, domesticating what is inconceivable. We are presenting a model here, sketching a bite-sized picture, using current models produced by scientific conceptualising. So our description may best be considered an approximation. It is a work-in-progress. No more.

All experiments result in the generation of data. The data produced by each multiverse incorporates information generated both during the experiment and at the experiment's end, when the final result is reviewed.

As each universe reaches the end of its intended purpose, the *observer* collects all the resulting information. However, during the course of each universe's cycles of expansion and contraction much information is also generated.

Effectively, each universe functions as a laboratory in which vast numbers of experiments of greater and lesser scale play out across space, from the sub-atomic to the intergalactic scales, and for various durations of time, from the micro-second to the aeon. Within these ranges countless experiments of diverse kinds are occurring. One is the experiment that involves organic life on Earth.

After this planet first formed, conditions conducive to the formation of biological life led to life emerging. An experiment involving biological life has played out ever since, utilising natural evolutionary and emergent processes. The human animal, Homo sapiens sapiens, is just one of untold species that have come into existence, settled into a niche, adapted to survive in changing environmental conditions, and either evolved into new species versions or died out. This same type of experiment involving biological life is repeated on countless planets throughout the universe. This planet is far from the only one.

Within the biological experiment currently taking place on Earth, each species may be viewed as a sub-experiment. So the human animal species is a sub-experiment within the larger biological experiment. This means that you, the reader of these words, are also an experiment. Before you were born you set certain specific psychological, social, intellectual and creative values in place, according to which you now live this life. The experiences you have, the lessons you draw, and the conclusions you come to, can be thought of as information drawn from the ongoing experiment that is your life. At your life's end you will acknowledge and process what resulted, extract further information from that result, and decide whether the experiment was successful, unsuccessful, or partially successful. You will then use this information to set the parameters for your next experiment, which will be your next human life.

This is exactly what the *observer* does in relation to each and every universe within the multiverse. Information is generated and collected throughout all levels of the kosmos. At the completion of each universe, the *observer* collects all the information.

Who, then, does the collecting during the course of each experiment? The answer to this question is spiritual identities like you. For the human, this occurs at the level of the node of Dao-consciousness.

CHAPTER 2

NODES OF DAO-CONSCIOUSNESS

WE BEGAN THE PREVIOUS CHAPTER by asserting that everything that exists derives from the Dao. This includes all spiritual identities, which we are identifying as nodes of Dao-consciousness. This is because, as we also stated, consciousness is intrinsic to the Dao and derives from it.

The Dao's nature is to manifest nodes. Nodes are individual agglomerations of consciousness that are cast from the Dao. By "cast from the Dao" we refer to the spontaneous natural development of a node that consists of the same substance as the Dao. Of course, the term *substance* is false, for the nature of the Dao has no substance in a conventional sense. So a node of Dao-consciousness may be thought of as a spontaneously agglomerated product of Dao-nature.

The arising of a node of Dao-consciousness is a spontaneous event triggered by a local accretion in density. A suitable metaphor is that of the ocean undergoing turbulent conditions, and as a result spontaneously emitting droplets from wave tips. This is a perfectly adequate metaphor, given that chaotic conditions in the sea generate forces sufficient to fragment the water, just as forces within the Dao generate something like tidal pressures. As a result, a variety of droplets come into being. Nodes of Dao-consciousness are these droplets. Like droplets, nodes consist of various magnitudes and complexity.

Not all nodes remain whole throughout their existence. Some nodes spontaneously fragment after their emergence from the Dao.

Different nodes fragment into different numbers of fragments. There is a basic correlation between the size of the node and the number of its fragments. Larger nodes fragment into more nodes, smaller nodes into fewer. This is true as a general statement, but there are exceptions. Fragmentation occurs in the case of nodes that co-associate with the human, as well as with the horse and the cetacean family, consisting of whales, dolphins and porpoises.

As regards numbers of fragments, nodes that co-associate with the human species fracture, on average, into one thousand individual fragments. This number is approximate. Some human-related nodes fracture into fewer fragments, some into more. Nodes that co-associate with horses and cetaceans fracture into fewer fragments than the human.

When first cast from the Dao the node has little self-awareness. It can be barely differentiated from the unmanifest from which it is freshly accreted. Given it is of the same nature, it feels associated with the unmanifest, to the extent that it sees little difference between itself and the tidal forces which led to its manifestation. It is still bobbing on the wave, so to speak, by which it was cast from the Dao. Development is required for it to begin to identify itself as distinct from the infinite unmanifest.

Gradually, the node becomes aware of its situation. It perceives others of like nature around it. It becomes motivated to find out more about where it is and who it is with. It learns from those who are like it that various dimensions of existence are available for exploration. It also learns that it is possible for it to eventually make its way back to the ultimate unmanifest from which it was cast. There is an underlay of excitement in learning this. It discovers it can explore. And it becomes eager to do so.

As it manifests its intent to explore, others respond. So it finds community. Within that community is information. But the information is not easily understood by the node. The node doesn't know enough, and it lacks the inner resources to comprehend what it instinctively knows is available and may be understood. It becomes aware of this because it can see that many of those it is with possess extensive knowledge, love and wisdom. It feels frustrated. So now, having developed a strong sense of its own limitations, and being motivated to extend itself beyond those limitations, it seeks opportunities to learn and grow.

Accordingly, it enquires regarding what specific opportunities are available to it to become loving and wise like others in its new-found community. It discovers appropriate opportunities have been mapped and thoroughly understood by those who possess greater experience. The community offers advice and counselling. This increases the node's awareness of what options may be preferred for self-development. By this means it is educated regarding what opportunities are most appropriate, and given the information it needs to make its first forays into learning. One such opportunity is to co-associate with biological species on this planet.

Environmental niches exist in various places across the Earth, for short or long periods. When a suitable environmental niche was identified from the spiritual level, nodes took responsibility for seeding it with appropriate organisms. Organisms were created, or modified from preexisting stock, and distributed in a way that breeding populations became established. They were then left to follow their own courses. Species that survived, did. Those that did not went extinct. There was never any particular concern over either outcome. It was the relationship between the characteristics of the niche and the surviving organism that mattered and was recorded.

This type of activity is part of an experimental imperative, which manifests as a creative drive, that all nodes of Dao-consciousness possess. The reason this creative drive was deliberately embedded in emergent nodes of Dao-consciousness was so the Dao itself could explore the consequences of physical life as it emerged, evolved, matured and came to the end of its cycle. And it enabled engendered nodes, which are of itself, to participate on all levels in the process. These levels are: outside, observing from a distance; nearby, participating by adjusting physical processes; and inside, directly experiencing all that occurred.

Originator, outside, near, inside, re-absorber—this is the mystery of existence that everyone and everything, from the smallest microbe to the largest Dao-entity, are collectively engaged in. The Dao is examining, experiencing, appreciating and extra-producing itself through us. Meanwhile the whole kosmos is shifting, fracturing, kaleidoscoping, reforming, and evolving. This is the wonder of what is!

Physicality enables nodes to test and stretch themselves in ways they cannot in the spiritual domain. Accordingly, almost all nodes choose to co-associate with bio-identities at some stage of their evolution. We won't say all do, because that is not the case. And there are varying degrees of involvement, which give rise to varying intensities of experience. There are also shorter or longer durations of co-association. All these parameters are selected by the nodes themselves.

The process of node fragments co-associating with the human species began not long after the development of the species Homo sapiens sapiens. Occupancy was prepared a long time before. Pre-human species, humanity's ancient forebears, were initially seeded by nodes from the spiritual level then left to follow their natural biological course. When they were ready to be occupied, that was done.

Occupancy began when it was observed that a sufficiently large breeding population had become established that those overseeing the process were confident it would survive as a viable species. Naturally, it took several million years to reach that point, given the entire process began with the initial evolution of biped animals, followed by their development into a distinct line of species. When, after observation, the decision was made that Homo sapiens sapiens was suitable for embodiment, and that the process of embodiment would be able to continue into the foreseeable future—which at that stage comprised the order of ten million years—the process of embodiment began.

The precise prognosis was that the species would allow not only the development, through continued embodiment, of individual history and karma, but there would also be sufficient time for all the embodied to resolve karma. Accordingly, an individual node of Dao-consciousness could be confident that specialising in this particular species would be adequate for the completion of its purpose, which is to achieve maturity and no longer require embodiment in an organic species. On the basis of this projection, the decision was made that the species was viable.

Each node of Dao-consciousness possesses all the qualities of the Dao. These qualities include identity, intellect and purpose. Consequently, each individual fragment of a node also possesses identity, intellect and purpose. So you who are reading this transmission are a fragment of a node of Dao-consciousness. You have individual identity, intellect and purpose. You have self-creativity and choice. You have utilised these qualities to shape all your prior incarnations for the purpose of experiencing and learning. And you will continue to utilise them to evolve according to your own creative intent.

Accordingly, it may be said that nodes of Dao-consciousness naturally seek to experience, learn and evolve. Entering physicality offers opportunities to do just that, which means nodes co-associate with physical species for their own advantage. Extending their awareness into the physical domain enables them to experience situations that are not available in the spiritual domain. For example, being in mortal fear of your life is not an experience that it is possible to have in the purely spiritual domain of the Dao. Nor is self-sacrifice, such as parents willingly do for their children, friends do for each other, or citizens do for their country. On the other hand, many other experiences do translate between the spiritual and physical domains, such as being creative, nurturing others, coming to understand others, and fostering environmental niches.

It could be said that just as the Dao's intent has created the multiverse as an extended experiment, setting up parameters then allowing them to play out in whichever way they do, so each node chooses a physical species to co-associate with, selects certain parameters that broadly define the nature of its possible experiences, then sets the co-association in motion to find what results.

In this sense, a node is intimately involved in its experiment. It could even be said that, given it is testing itself, it *is* the experiment. The node has a vested interest in the experiment because it evolves according to how the experiment plays out. To the extent that the Dao eventually receives everything nodes experience, it could be said that the Dao equally has a vested interest in the result of each node's self-experimentations, as it also evolves according to the result.

A node fragment of Dao-consciousness, being an individual spiritual identity, first associates with a human body when the body is an embryo in its mother's womb. [*Editor: A particular embryo is chosen for reasons outlined in the next chapter.*] An intention to co-associate is then directed towards the embryo. Beyond this there are no particular levers to pull or any kinds of biological or energetic structures to manipulate. Simple intention is sufficient to ensure the spiritual identity and its chosen body are united for the duration of that body's life.

The blending phase requires the insertion of a point of attention. A node fragment has only to focus its intention, then sustain that intention, for there to be effective co-association between the spiritual Dao-identity and its selected physical bio-identity. This is all that is required. That is why it is called merging, coalescence or co-association. One merges into the other and a new composite identity, a bi-located identity comprised of Dao-consciousness and bio-consciousness, comes into existence. This occurs while the foetus is still in its mother's womb.

How does co-association take place? To speak in an energetic sense, the node fragment of Dao-consciousness that locates itself beside, within and through the human body exists in the shape of a sphere. Its structure is globular. Within the globular spiritual structure, at its centre, is what might be thought of as a kernel. This is the centre from which filaments radiate. The filaments contain information that the node fragment carries with it from and into each life. The kernel could be conceived of as being somewhat like a small seed within a translucent globe covered and filled with patterned filaments of light. But, of course, the kernel exists electrospiritually, not physically.

When a spiritual identity seeks to co-associate with a body, it brings the centre of its globular structure into alignment with a central energetic point within the body's aura. This energetic point becomes the hara dantian, just below the belly button. In this way the centre of the spiritual globular structure is aligned with the centre of the physical body. Intention then sustains this alignment for the duration of the life.

Everything requires preparation. The inexperienced node fragment needs to learn how to align its kernel centre with the aura's hara centre in order to sustain its connection with a physical body. Only after learning to do this can it begin co-associating with the body of any species.

To conclude this description, and to make what we have said clear, no binding or locking into position is required for the spiritual self and the human bodily self to be aligned. When there is an intention to merge, merging occurs. When there is intention to dissociate, merging is terminated. There is complete freedom for the identity to leave its association with the body for a time, which is naturally done during some phases of sleep, or to sever the connection completely—even, if it chooses and for whatever reason, before the natural death of the body.

Learning to become a balanced bi-located identity, in which Dao-identity openly and appropriately contributes to life experience alongside bio-identity, is an essential aspect of what is involved when a node fragment utilises human experience to achieve maturity.

One last point needs to be made in relation to nodes and fragmentation. After all a node's fragments have matured sufficiently, they are designated as having completed their cycles of reincarnation. They then reunite with every other fragment from the same node. In doing so, each one brings back everything they have experienced and learned, every skill they have developed, the wisdom generated from every responsibility they have taken on and successfully negotiated. As a result, the node is incalculably enriched in comparison to its initial inexperienced state.

Underlying any and each exploration is the drive for enrichment. Enrichment of the individual. Enrichment of whatever environment a fragment occupies to which it can contribute. Enrichment of the node when the stage of reunification is reached. And enrichment at other levels beyond that of the reunified node. The process of enrichment goes all the way back to the Dao, to which everything is eventually returned.

The image of a node emerging from the Dao, spontaneously fragmenting, and its thousand or so fragments then independently spreading out through the kosmos to explore, experience, interact, learn and evolve, could sound to some like a process a writer of science fiction might concoct. Nonetheless, it is true.

The physical multiverse is your domain. The expansive dimensions of spiritual space are your domain. Embodied realms, non-embodied realms—they all present opportunities for inquisitive nodes and their questing fragments. You are currently bound to a physical form. And you have definitely made a commitment to co-associate for a time with the human species in order to work through the conditions and experiments you yourself have set in train. But this will not always be so. When you take your place within a reunified node of Dao-consciousness, the kosmos in all its variety and vastness will await your engagement. The experiment is that vast and that fascinating. And it is as exciting as you choose to make it.

CHAPTER 3

EACH LIFE IS AN EXPERIMENT

FROM THE SPIRITUAL PERSPECTIVE, EACH LIFE, each incursion by a node of Dao-consciousness into the physical realm, each occasion a node fragment enters a human body and its associated social environment and experiences a human life, is an experiment in consciousness.

Every node fragment experiences itself as an individual spiritual identity. Each occasion it co-associates with a body, and each life it consequently lives, is an experiment in which certain very specific parameters have been set in place prior to birth. And the experiment is to discover what the spiritual node fragment identity, experiencing the human domain from within that body, makes of the opportunity. From this perspective, each life is an experiment in which an ongoing spiritual identity explores a particular aspect of human existence in order to enrich its own consciousness—given that whatever the spiritual identity experiences, processes and learns during the course of each life goes back to its ongoing identity at the node fragment level, becoming material that feeds its enrichment and so its evolution.

Reincarnation is fundamental to this process. This is because the purpose and process of living human existence and the purpose and process of each fragment's existence as a spiritual being are equivalent. They are equivalent because human life has been organised to provide the experiences spiritual identities need to learn and evolve.

Every child goes through the process of learning to crawl, to walk, to talk, how to go to the toilet, and how to interact acceptably with others. Every child goes to school to learn practical, creative, social and intellectual skills, the aim being that they may eventually take their place as an adult who acts effectively and responsibly in the human world. This exact same process is reflected in the spiritual dimension. In fact, the socialisation of a child from naive and unskilled to experienced, knowing and self-reliant, who is also able to feel compassionately for others, is directly reflected on the spiritual level.

Every node fragment begins its existence as naive, unskilled and unknowing. Every embodied node fragment initially lacks the inner depth needed to appreciate and sympathise with what others around it are going through. A child doesn't appreciate what adults around it are feeling and thinking because it lacks adult experience. Each child progresses through a variety of experiences, from child to teen to adult, which happens naturally as they age. As individuals graduate from the education system and enter the workforce they inevitably interact with a range of other people. Some they lock horns with, others they enjoy playing with, yet others they work happily or unhappily with. In the process everyone learns about themselves and others. This is the normal progression of human learning. The key to learning, however, is not memorising facts or passing exams. The key to learning is having a range of diverse experiences, drawing lessons from them in order to perform in the future with greater perception and skills, and to apply them in ever-widening spheres of human activity.

Eventually, the individual becomes knowledgeable in many aspects of human existence, develops expertise in a range of human activities, learns to act out of love in all situations, and manifests a combination of deep qualities that lead others to say of that individual that he or she is wise. Clearly, this higher level of human development does not occur in one life. It takes many lives. That is why repeated lives need to be lived, and why reincarnation is axiomatic to each node fragment's incursion into human existence.

From this it can be seen that incarnating on this planet is a long-term commitment. It is a testing process. It requires application and effort. The pay-off is that what is experienced and learned, the skills that are developed, the expertise that is acquired, the capacity to love that blossoms, and the wisdom that eventually infuses the spiritual identity as it participates in the complexities of human existence, all feed directly into the node fragment. As a result, it progressively, life by life, becomes enriched and develops into an identity that is far greater than it was previously, when still a naive and inexperienced fragment. This, in a nutshell, is the purpose of reincarnation. It provides a mechanism by which naive and inexperienced spiritual identities may learn to become skilled, knowing and loving.

Learning requires planning. A child does not go to school each day to have random encounters. Schools have teachers, and teachers have lesson plans. Lessons are designed to progressively impart information and enable students to develop skills, with each new lesson building on what has gone before. The same process applies to reincarnation. It requires planning. Lessons are organised sequentially. And skills and information are imparted as and when they are needed so the individual may take the next step in its development.

However, there is a fundamental difference between a child's schooling and what a spiritual identity learns during the course of repeated incarnations. A child has a teacher, and the teacher has lesson plans that students are required to follow. This does not apply to reincarnation. Instead of there being a teacher, each spiritual identity is its own teacher. And instead of prescribed class lessons, each spiritual identity formulates its own life plans.

This means that each life is not an experiment anyone else is performing on you, but is an experiment you yourself are carrying out, over an extended sequence of lives, as you explore the numerous possible activities, roles and relationships available in the human realm. As a consequence, the process of entering a body is not haphazard. Before incarnation each identity decides on principal tasks and goals, selects psychological traits that will facilitate achieving them, and generates a life plan that may, or may not, be followed during the course of living the chosen life. There are reasons for both outcomes.

Life plans vary considerably. Some are simple, others are complex. In some lives individuals take on many tasks, in others few. Some lives are self-confrontational, being designed to help individuals confront, challenge and test themselves. Other lives are planned to be serene. In each life, situations are planned so that lessons may be learned. Some lessons are simple and straightforward, others deep and complex. Some lessons are easily learned in one life. Other lessons need to be repeated, life after life, in a series of different situations, in order to grapple with all the nuances involved, and so that many related skills are developed and all interlaced lessons extracted.

From this it may be seen that no life plan is the same. For each incarnating node fragment every life has a unique life plan. Given node fragments incarnate an average of one thousand times, that means each will have one thousand unique life plans. Nonetheless, there are a number of parameters fundamental to each life plan that are considered during pre-life planning. These include:

- Selecting a body, with its particular genetic disposition and inherited physical and psychophysical traits.
- Selecting the cultural conditions, including social norms and belief systems, into which the individual will be born and which will initially shape the individual's outlook.
- Selecting skills on which to focus during the upcoming life. In general, skills are chosen because they build on what the individual has developed in previous lives or fill in gaps in skills the individual requires to develop.
- Selecting psychological traits, both positive and negative, from those that the identity has generated during the course of previous incarnations.
- Identifying karmic issues, again drawn from previous incarnations, that will be addressed during the next incarnation.
- Selecting key tasks that address karmic issues and promote growth.

- Selecting obstacles, consisting of situations and particular people, that require extra effort to be overcome and so assist in the accomplishment of major aspects of the life plan.
- Making agreements to meet and be assisted by other individuals, and agreeing to assist others to achieve their own aims.
- Organising the conditions for all these various tasks and key choices to occur.

The life plan weaves all these together into a complex mesh. Ultimately, the purpose behind each life plan is to facilitate the spiritual identity's ongoing growth. What adds further complexity to this process is that what happens in a life may or may not fully implement what was intended. Lives may go off plan, a little or a lot. Sometimes deviation is a matter of choice, because despite having a plan individuals retain choice always and in all circumstances. Sometimes deviations occur due to factors beyond the individual's control, such as being involved in a natural disaster, or having an accident, or suffering a random act of violence.

Accordingly, while each life has its plan, the extent to which a life adheres to that plan will never be known until the life is lived. Nonetheless, whatever happens provides experiences, the opportunity to learn lessons, and stimulates further growth. In this way each life, and each sequence of lives, becomes a complex series of experiments, the details of which are planned, but the outcomes of which are never known until the life is complete and the individual reviews what occurred.

Sometimes, when a spiritual identity reviews a life after its completion, the life is considered disappointing, because conditions set in place prior to the individual being born were not explored to the extent that was intended. Pre-organised opportunities were not taken up. Yet at other times expectations are exceeded, because more insight and understanding is achieved than was anticipated.

In practice, the outcome of a life is almost never completely what was anticipated. Indeed, a sizeable percentage of lives do not end up at all resembling what was anticipated prior to embodiment. Yet this lack of predictability is not a negative. It is just how it is. It is what makes each and every life journey an experiment. An experiment in experiencing, processing, responding, learning and growing.

A single human life may be called an experiment because certain pre-conditions were set in place prior to birth, and the aim is always to discover how the spiritual identity copes with the resulting circumstances and uses them to navigate through life.

We deliberately use the term *navigate* here because, as we just explained, while a life's parameters are pre-determined, once the life journey begins it is over to the individual to navigate through and around all the obstacles, opportunities and distractions it encounters. The end result of a life is not adjudged a success or failure on the basis of what it achieved externally, although some lives certainly do have a physical outcome as their goal. Rather success and failure are assessed on the basis of what was experienced, what was learned from experiences, and to what extent the individual used them to enrich itself and evolve.

Life outcomes are always useful and interesting, no matter whether they adhere closely to the life plan or depart wildly from it. A scientist may set up an experiment and, once the results are in, exclaim, "I didn't expect that to happen!" Yet he may actually be pleased because the unexpected result has opened up a whole new field of research. Similarly, on reviewing a just-lived life, an individual may see that little panned out as expected, yet what was experienced and learned has opened whole new possibilities for future experimentation. In all this we need to make clear that there is no big scientist in the sky who is experimenting with your embodiment in the human domain. Rather, it is you yourself who is both the subject of the experiment and doing the experimenting.

This is why we say that becoming embodied as a human being is an experiment in consciousness. Life after life you are testing yourself in the cauldron of human existence. The experiment in incarnation is completed when you have learned to love selflessly and gained mastery at being a node fragment living in a body.

Billions of node fragments of Dao-consciousness have been seeded into billions of human bodies, which themselves are planted into the soils of different physical and social environments, and fed very different kinds and intensities of inputs. As a consequence, each undergoes different combinations of experiences during the course of each life.

Some of these experiences are chosen, some occur by accident, some result from uncontrolled natural forces, and some are impacted by others' malevolent intent. Some individuals live for a brief period, others for a century. Some live in traumatic environments, others live placid lives. Some intensely explore very specific areas of human activity and develop expertise in those areas. Others make little effort. Yet others are thwarted from day one, and either end up frustrated and embittered or learn acceptance and to go with the flow. Some make one mistake and pay for it for the rest of their life. Others live charmed lives and escape all consequences of the various scrapes they get into. And no matter what the life experience, everything is processed by an individual personality in which different traits predominate. So the ways individuals perceive, process and respond to the same experiences, let alone to vastly different experiences, varies enormously.

This is why human experience is so richly nuanced. And why each life is a pre-planned experiment for which the outcome is never fixed. However, for all node fragments one same result *does* occur: they accumulate multiplexed layers of human experience and identity.

CHAPTER 4

ACCUMULATING HUMAN IDENTITY

AS A NODE FRAGMENT PROGRESSES through the trajectory of its lives, it comes to appreciate what works best for it in the human domain. What works best is learned through trial and error. Some latch on to what works for them quickly. For others it takes much longer. Some are naturally more conservative and play it safer in their choices, while others like to stretch themselves, perhaps even to breaking point—of the human being they have incarnated into, of course, not of themselves as a node fragment.

However the node fragment comes to appreciate what works best for it, and whether that appreciation comes earlier or later, the result is that it accumulates favoured ways of functioning in the human domain. This results in the development of accumulated human identity.

Accumulated human identity comprises all the character traits a node fragment has "tried on for size" throughout its lives. Some traits are found conducive to growth, or are just plain fun to manifest and so are fostered, while others the node fragment does not find useful or enjoyable and chooses not to continue working with. Some character traits lead to benevolent behaviour while others create problems for the individual and for those with whom it interacts. These latter are obstacle traits, which play a very useful role in each individual's growth.

This means that the character traits present in a node fragment's accumulated human identity are derived from personal experiences undergone life after life. But these traits are themselves part of an even larger store of experiential material, which includes: experiences resulting from choices made, neglected or rejected; repeated behaviours, changed behaviours, repeated reactions to others' behaviours; successes, mistakes and lessons learned; and skills, abilities and talents developed in the many different fields of human endeavour. There is also a store of karma, a store of obligations (which results from as yet unfulfilled promises made to reciprocate help the node fragment has received), and a store of goodwill. This last incorporates the largesse of compassion, benevolence, and love that the node fragment naturally shares in its non-embodied state but that it has to learn to share in the human domain.

Listed baldly in this way, it is a daunting store of experiential material. It indicates why, as experience accrues from hundreds of incarnations, choosing the components of future lives becomes ever more complex. Much experience has been accumulated, but the individual has not yet learned how to juggle its preferred character traits, its stores of skills, its long-term aims, its karmic responsibilities, its obligations, and its desire to share, with the daily pleasant and trying situations basic to human existence, and its own often inappropriate responses to those situations, which lead to transgressions that then need to be addressed in later lives. All these factors contribute to the shaping of each individual's accumulated human identity and so need to be taken into account when a new life plan is being formulated.

To explain the complexity involved, we offer the model of the five-layered self.

The model of the five-layered self incorporates five levels of being: the biological self, the socialized self, the essence self, the energetic self and the spiritual self.

The *biological self*, more commonly known as the body, is grounded in the human animal, Homo sapiens sapiens. The central nervous system, the limbic system and the brain's cognitive capacities collectively facilitate the receiving, processing and passing on of complex information. The nature of the biological self is fundamentally dictated by genetically inherited traits. However, influences external to the DNA affect the expression of genes during the body's growth. For example, famine will alter the body's inherited features. Deprived of vital minerals and trace elements, the body's growth is stunted. Equally, physical, emotional and sexual trauma will disrupt important connections throughout the body, limbic system and brain and prevent normal development.

The *socialized self* is a psychological and behavioural layer that is shaped by the physical and social conditions into which the body is born and that influence its development. Language, education, family, social norms, work, opportunities, and so on, all contribute to the formation of the socialised self. So where the biological self is shaped genetically, the socialised self is shaped by social and environmental conditions and by the way the individual responds to those conditions. Given people instinctively feel their identity is their body, the biological self and the socialised self are bound up in each other and function in tandem.

The *essence self* consists of the higher human capacities. It is where human beings grow practically, emotionally and intellectually. It is the essence self that is educated in the fullest sense and that learns, develops abilities and matures as an individual. Yet the essence self needs to use a particular language and socially conditioned norms and outlooks to live and express itself. So it interacts with the external world via the socialised and biological selves.

The biological self is largely mature by the age of eighteen. In contrast, the socialised self is labelled as adult at a symbolic age, which varies in different cultures from fourteen to twenty-one. In fact, the socialised self never matures. This is because it is formed as a negotiation between the biological and essence selves and the social environment. And it only ever changes as much or as little as is required to maintain its ongoing body-based identity.

The *energetic self* stands between the three layers of the biological, socialized and essence selves on one side and the spiritual self on the other. Those three layers in themselves deliver rich experiences and provide much material for an individual to work with. But for those who wish to appreciate their spiritual nature, an understanding of the energetic self is crucial. The energetic self is part of an individual's electromagnetic functioning. It extends into and blends with the globular structure of the node fragment of Dao-consciousness that is co-associating with the occupied body. As such, the energetic self incorporates the energy nodes known as chakras as well as the electrospiritual aura. The aura plays a significant role because it conveys information between the spiritual domain occupied by the individual's node fragment and the physical domain occupied by the biological, socialised and essence selves.

The *spiritual self* is an incarnated individual's core consciousness. It is a portion of a node fragment's spiritual identity. The spiritual self is the principal conduit of cues that are sent by the spiritual identity into its embodied manifestation in order to ensure the life plan is adhered to. That the plan is not adhered to is often because the spiritual self lacks direct input into the embodied everyday awareness, and so is unable to contribute to the choices the individual makes. One of the ultimate aims of embodiment is to learn how to establish a connection between identity at the node fragment level and identity at the embodied level. The embedded spiritual self is the conduit for such a connection.

Each of the biological, socialised, essence and energetic selves exist only for the duration of a particular life. They disintegrate when the body dies. At that moment the spiritual self leaves the body and returns to the node fragment level—although, technically, it has never left. This is an arcane issue that is too complex to discuss here.

What can be said is that before the biological, socialised, essence and energetic selves disintegrate all that occurred within each layer is uploaded to the identity at the node fragment level and stored as part of accumulated human identity. Relating this process to the model of a node fragment as a sphere, the uploaded information is stored in the filaments that extend from the kernel at the centre of the globular structure to its peripheral boundaries.

In this way, nothing that occurs in a life is lost. The uploaded information consists of experience, so it is extremely dense. Everything, good, bad and indifferent, is uploaded to the Dao-consciousness level. After the individual has recovered from the life and got back its breath, so to speak, it uses this newly uploaded material to review the life, examine the choices made or not made, ascertain why this occurred, extract life lessons, and use those lessons to shape its future life plans, building on what was well done and correcting what was not. This is how everything from each life contributes to a node fragment's ongoing development.

The most significant factor that prevents individuals from fulfilling their life plans is fear. Fear is normally viewed as an instinctive reaction to physical threats. Fear on this level can stop individuals from engaging with experiences that are part of their life plan. However, there is a much deeper, more pervasive level of fear that causes much greater deviance from life plans and therefore much deeper frustration. This fear is buried deep in each individual's psychology.

Once a node fragment selects a particular body, and that body is born, it emerges into a social environment that is not within its control. During the infant's first two years it has no ability to control the body in which it resides. All the individual can do is ride out the experience as best it can. To do so it develops behavioural coping mechanisms. These mechanisms help the developing identity get along with others. In most cases this begins with close family. Because the family situation is chosen prior to incarnation, and because the particular psychological traits the infant has at its disposal are also chosen, this means the coping mechanisms are not present accidentally. They are part of the childhood experiences node fragments select before incarnating.

Children's experiences vary enormously. However, all are impacted by the ways they are nurtured, pressured and, in many cases, traumatised. Coping mechanisms develop to help the individual negotiate all three. In this way coping mechanisms are fundamental to the socialised self's development. All coping mechanisms develop in response to uncomfortable and threatening social situations. To this extent they have a socialised fear pulsing at their psychological core. Examples include the fear of being vulnerable, the fear of losing control, and the fear of being left out. As childhood progresses into the teen years, an individual's earliest coping mechanisms become buried deeper and deeper within the individual's psychological make-up, being covered as new coping mechanisms are adopted. The earliest coping behaviours are still present. They underlie all behaviour, but are acted on unconsciously. Effectively, they are the socialised self's underlying drivers, being deeply embedded traits to which individuals become so habituated that they are considered to constitute who they are.

Buried fear and its related coping mechanisms constitute a psychological construct the psychiatrist Carl Jung called the shadow self. It incorporates all the negative traits that lurk in the basement of the self. They are in the basement because most people shut the door on them and try to pretend they are not there. But they are. And sometimes stress causes the door to burst open and they leap out. The result is that individuals behave negatively in ways that are not typical of them. When this occurs others observe they are not their normal self. In fact, they *are* their normal self, it is just that it has become so deeply buried no one usually sees this self, including the person whose behaviour is supposedly abnormal.

There is more to this situation. This is that even though the shadow self lurking in the basement mostly stays there, with the door firmly closed on it so it cannot manifest in everyday situations, a "smell" still emanates from the basement. This "smell" flavours the individual's socialised self, with the result that its behaviour is subliminally influenced by the shadow self's traits. These buried behaviours become obstacles individuals must overcome to fulfil their life plan.

In general human beings do not learn as much when life goes easily and well. They learn best when they come up against problems and difficulties. When spiritual identities first begin incarnating in a human body they find the experience overwhelming. Human life is complex. Being human is difficult. It takes many incarnations to learn to steer the body around, to cope with hormonal tides, to negotiate human social interactions, and to develop personal essence abilities. It takes even longer to learn how to operate all five layers of the self in concert.

For a long time incarnated individuals find other people difficult to figure out. They find themselves difficult to figure out. Others do things they don't understand, for reasons they don't comprehend. They themselves do things they don't understand, for reasons they don't comprehend. In addition, daily life is full of obstacles which stop individuals in their tracks. Practical obstacles come in the form of financial and time pressures, lack of opportunity, lack of ability. Emotional obstacles take the form of self-inhabiting character traits, such as laziness, impatience, self-pity, resentment, or a tendency to addiction or violence. Intellectual obstacles include an inability to think straight, a refusal to acknowledge reality, and resorting to prejudice.

Obstacles also manifest in the form of individuals who perform the role of antagonist in another's life, getting in their way and, from the affected individual's perspective, preventing them from achieving their goals. In many cases there is a karmic relationship between protagonist and antagonist, with unresolved relationship issues hanging between them. However, this is not always the case.

Karma itself is widely misunderstood. A common belief is that if a person is good in one life they come back in a good situation in their next life. Alternatively, if they are bad then they are born into a negative situation. This view of karma is grounded in a reward and punishment scenario. As a result, people talk of good karma and bad karma. So if you are helpful to another that is supposed to result in the accrual of good karma, and visa versa. This is much too simplistic. It is wrong.

Karmic relationships exist mainly as a result of psychological impinging. So addressing karmic debts involves resolving interpersonal behaviour. As a result, an individual isn't born into a tough life situation as punishment, but because it provides the obstacle conditions they have decided they require to get them to address their own psychological limitations. That an antagonist is also present puts pressure on the individual to face up to shadow traits, rather than keep them locked in the basement, given avoidance is easy if no one is getting in their face and forcing them to look at themselves and their failings. This means that for each person obstacles have been organised to arrive at a certain stage in their life in order to provide a specific opportunity. Sometimes these obstacles are so well integrated into daily life that they appear as just another problem to be dealt with. At other times the obstacle arrives from left field and creates a huge disturbance, perhaps sending individuals in a wholly new direction that they had never previously contemplated. Such is the surprise and joy, or discomfort, even horror, of incarnated existence.

The overall purpose in incorporating obstacles into a life plan is to address the fact that each individual's accumulated human identity contains both positive and negative traits. By drawing negative traits down into a life, they can be addressed and their deep underlying psychological causes faced up to and eliminated. In doing so, the accumulated human identity simultaneously builds up its positive abilities through confronting and dealing with negative traits, and purifies itself of those traits. Eventually, after much effort, it develops into an experienced, multi-skilled, knowledgeable, compassionate, loving and wise node fragment of Dao-consciousness.

CHAPTER 5

LEARNING WHO YOU ARE

THREE CONDITIONS ENSURE THAT WHEN a node fragment of Dao-consciousness enters a body and occupies it for a life, the experience is vivid and engaging. These conditions are identification, attachment and forgetfulness.

Identification experientially embeds the node fragment's awareness in the body it is occupying. So when the body experiences hunger, pain, pleasure or ecstasy, the node fragment itself experiences hunger, pain, pleasure and ecstasy. Attachment to others generates loving relationships, non-loving relationships, and the raft of possibilities between, thereby locking the fragment's awareness into the human world. And forgetfulness keeps each life an independent experience. If a node fragment carried all its memories of prior encounters into a life, the current life experience would be overwhelmed by them, preventing a fresh life from being experienced and lived. Forgetfulness ensures individuals are not encumbered by reactions to prior incidents.

However, there is a disadvantage in not remembering and in being locked in a state of identification and attachment. This is that the embodied individual forgets who they are, that at their core they are a spiritual identity, and that they are here, on this planet, living in a body and dealing with particular life experiences, because that is what they themselves chose to do. It is certainly the case that when an individual knows who they are, in the sense of what qualities they have drawn on from their accumulated human identity for this life, and the nature of the tasks they have set themselves this time round, the life becomes more meaningful and many frustrations vanish.

Frustration is a by-product of being identified, attached and forgetting prior existences, including between life existence in a wholly spiritual state. Frustration occurs because deep inside everyone has a spiritual self, and it knows that its natural state is not being encumbered by a body but being free. After an extended series of incarnations, this frustration begins leaking out of the spiritual self into the embodied individual's everyday awareness. As a result, the embodied individual starts wanting to know more about itself, who it really is, where it came from, and how it came to be here. In other words, the individual starts enquiring into the nature of its life and self. For many this extends to enquiring into the nature of life in general, and how the universe is put together and works the way it does.

There is much contention today about where this desire to enquire comes from. Many maintain it is a product of biology, that having a nimble mind that wants to delve into the world gives human beings an evolutionary advantage. We disagree. We see the desire to lift up the curtain, so to speak, and peer into the working of reality, is a function of the spiritual self. The scientist's urge to delve into the workings of the world, the artist's desire to express their vision of the world, the horticulturalist's urge to creatively nurture a garden—these are all by-products of the spiritual self's frustration with living in darkness, not knowing what is really going on, and wanting to perceive beyond the surface of life experience and explore the depths.

This is why the urge to learn who you are is such a significant step. It signals the arrival into your everyday awareness of the push from your spiritual self to understand you are a node fragment, which has sent part of itself to co-associate with a body and live a life. One of the goals of the incarnational cycle is to increasingly merge your spiritual self with your everyday self. Doing so also necessarily involves remembering key aspects of past lives as they impact on and have shaped your current life. Feeling frustrated, and acting on that frustration by delving into the who, what, where and why of your existence, is a reflection of growing maturity. It takes a significant number of lives before an individual becomes ready to delve into the workings of their current life as they are living it. The urge to delve occurs during a particular phase in each individual's overall incarnational cycle.

The first phase of incarnation involves overcoming the trauma of incarnation, of no longer being overwhelmed by all the biological and social inputs impacting on one's everyday awareness.

The second phase of incarnation involves developing confidence in oneself. This manifests in beginning to be comfortable as a spiritual identity living inside a body—even though at this stage one is still not consciously aware this is the case. This phase involves starting to put feelers out, to try doing different things, to interact more openly with others, and to assert a little personal order onto the chaotic impressions, feelings and thoughts that characterised experiences during the first phase. In this second phase one starts developing rudimentary control over what input enters the awareness as well as what one gives out in the form of words and actions.

During the third phase one feels very comfortable with the process of incarnation, to the degree that one asserts one's identity in the world, sets goals and does what is needed to achieve them, and organises others so one may do as one desires, given that organising variously involves confronting, avoiding, overcoming, persuading, forcing, coercing, diverting or ignoring compliant or noncompliant others.

These three phases of incarnation may be likened to infant, child and teenager. During the infant phase one is largely helpless in the face of one's experience of being incarnated. During the child phase, having by now experienced incarnation a large number of times, one is starting to realise not everything is a threat and that one can interact with others in a fruitful way. During the teen phase one is involved in forging a self that can act as an independent identity in the world. These three phases are reflected in the growing breadth and depth of an individual's accumulated human identity.

As a broad statement, it could be said that the infant phase starts out traumatic, but incarnation gets easier to cope with as one has more incarnations and so becomes familiar with what is involved. Nonetheless, one is aware, not consciously but at a deep level, that one is dependent on others for survival. Many at this stage in the incarnation cycle take refuge in small communities or tribes. As a result, identity comes from being part of a shared social environment.

The child phase in some senses is the easiest. Individuals at this stage in their cycle of incarnations are still dependent on a social environment for their identity but are quite happy to comply with what that environment requires of them. At this stage such individuals are potty trained, so to speak, and can look after themselves with a modicum of independence. But they still need a supportive social environment to work within in order to feel safe and loved.

During the teen phase incarnation gets bumpy. Just as teenagers have to challenge their parents and themselves in order to break free of the home nest, so individuals at this phase of their cycles need to break free of the supportive social networks they previously relied on. Accordingly, they now set themselves more challenging tasks to carry out during each life. Just like the teen years in human society, the teen phase of incarnation is full of both confident and troubled individuals who are striving to make their mark in the world. Many times they screw up. But between lives they review what they sought to achieve, analyse what went wrong, and generate a life plan for their next life in order to do better next time. Life plans during the teen phase are more complex and more demanding than during the two prior phases, with greater results if they succeed, but correspondingly carrying a greater risk of failure. In general, most karmic impinging that arises during this phase is of a thoughtless variety, resulting from lack of understanding of consequences and the impact one's choices have on others.

This brings us to the adult phase of the cycle of incarnations. As with adults in the human world, this phase is about acting as a morally autonomous being and taking responsibility for one's actions. The adult phase is the longest. Just as human adult years nominally extend from twenty-one to death, so the complexity involved in growing from an immature adult spiritual identity to a mature adult spiritual identity involves a huge amount of experiencing, interacting, processing, learning and trying again. So this phase is extremely varied and, at times, can become somewhat murky. Due to the many things going on in each life, it is always complex.

The adult phase is primarily the phase to which we are directing the information in this and all the other communications we have initiated over the past several decades. Our purpose is to promote growth from the immature adult to the mature adult. Naturally, this occurs over an extended series of lifetimes. Nonetheless, the more one takes responsibility for one's actions the more quickly one can traverse this phase of the incarnation cycle.

Much has to be learned in order to become skilful, loving and wise. If everything individuals need to learn was released to them all at once, they would be overwhelmed and feel unable to cope with it. They would give up rather than enquire and delve. So this is not done. Instead, information is drip-fed progressively into incarnated individuals' everyday awareness so they can progress step by step. The speed that information becomes available is dependent on the pace of the individual's learning. As more information is required in order to make sense of what is happening, so it is released. This means that each individual sets both the agenda and pace of their progress.

The source of the information is primarily the individual's spiritual self embedded in their five-layered self. However, external sources usually have more impact during individuals' initial enquiries. External sources include books, other knowledgeable people who have adopted teaching and sharing roles, and courses and workshops. External sources are useful when they provide information and practices that the individual can apply to their own situation. However, delving deep necessarily involves seeking information within, information to which the spiritual self has access. In terms of becoming aware of one's life plan, and learning what and why specific positive and negative traits are present in this life, that information is available at the node fragment level. This is because what is going on in this life reflects what is present in the accumulated human identity. In a very real sense, an individual's identity in this life is an extension of what exists at that level. So this life has its roots in lives past.

Elsewhere the model of multipersonhood has been proposed to explain the nature of multi-life identity. This is a useful concept. Who you are now is a function of who you have been in prior lives, and what choices you made, what karma you accrued, what abilities you developed, and what negative traits you adopted in the process.

So if, for example, you have a fear of being vulnerable lurking in the basement of your current personality, it is not there randomly. You have selected it because it is useful for you to do so. It enables you to deal with specific issues you have chosen as part of your current life plan. In order to delve into your fear of being vulnerable, you ultimately need to go back to past life experiences. Resolving negative psychological scenarios doesn't involve just going back to examine your childhood experiences. Those particular childhood experiences were organised by you so you could address them and overcome their psychological impact on you. But they were chosen because they embody unresolved issues from previous lives. So getting to the deep roots of your fear of being vulnerable in this life would necessarily involve remembering what occurred in past lives.

In saying this, we are not suggesting that any current negative and self-limiting issues cannot be resolved unless their prior life arising is understood. Mostly the past life roots are addressed between lives, and the current life is deliberately organised so past life scenarios will be repeated this time round. This means that when you address an issue in this life, in effect you are also addressing what occurred in a prior life. As a result, you do not need to be aware of past life incidents in order to progress in this life. Nonetheless, remembering what happened, and appreciating how what once happened is impacting on how you feel, think and behave this time round, can be very clarifying.

Ultimately, the information you receive is on a need to know basis. As you need to know something, your spiritual self will make it available to your everyday mind. In order to get the most of your spiritual self's knowledge, it is necessary to make yourself available. You have to learn to open up your awareness and receive communications that, compared to everyday communications, are subtle and quiet.

Meditation is certainly useful in this regard. Meditation helps quieten your normally bubbling mind, still your thoughts, and open your awareness up to what we call inner cues. These are communications from your spiritual self that provide information on significant occasions that cue you to what is needed to fulfil the next phase of your life plan. However, not everyone is comfortable meditating. Another way communication is shared is via dreams. For this reason keeping a journal and recording dreams is a useful exercise.

To conclude, we are advising those who wish to learn who they are that they adopt a multiperson model. See yourself not as an isolated identity, but imagine a long line of identities stand behind you, involving hundreds of identities, extending into the distant past, as well as into the yet-to-be lived future. This will give you a more accurate notion of who you are and where you stand in relation to your node fragment's still accumulating human identity. The way to appreciate all this is by engaging in an extended period of self-enquiry.

CHAPTER 6

THE PROCESS OF SELF-ENQUIRY

WHEN YOU ARE BORN NO ONE KNOWS where precisely your life will lead. There is a plan, but no one either on Earth or in the spiritual realm knows how much of the plan will be realised in a particular life. That is the reason each life keeps everyone involved on their toes. You only find out what will happen during the course of your life by living your life. That is the wonder of incarnation. That is its mystery.

It is because people are aware a mystery beats at the heart of their life that they question their existence. We say this because, like you, we have lived many lives and so are speaking from experience when we say it is common for people to feel that a veil hides the deepest aspects of their life. Often this feeling includes the intuition that some kind of plan is in place. People want to be assured there's a reason they've been born at this time, into this body and this family, and into the particular circumstances that govern their life. The reason people ask deep questions is to get to the bottom of it all, to lift the veil and penetrate the mystery hidden at the centre of their life.

In response to this desire people have historically adopted a range of strategies. Many participated in a religion. However, religions offer only very generalised statements on the nature of life. Those seeking more personalised insights had to go elsewhere. For significant numbers this involved using divinatory tools, such as the I Ching, or reading signs in teacups or in the fall of sticks. Today people consult experts in psychotherapy to understand more about themselves and to face up to self-limiting traits. Many more consult mediums, although these are generally utilised to ease emotional pain and help make decisions rather than to delve into deep matters.

From our perspective, the weakness in all these is that they are somewhat haphazard. Valid information may be gleaned through them. And the spiritual self may, at times, utilise any one to share deep information with the enquiring individual. However, much more useful and sustained results are obtained when individuals adopt a consistently followed line of investigation. We call such an investigation self-enquiry.

At this point it is useful to step back and remind our readers that self-enquiry extends beyond embodied experience on this planet. Deep within you is an identity that has come from the spiritual domain. When this life is over, it, that is, you, will return there. So while on one level the purpose of self-enquiry is to delve into the psychological workings of your current existence and uncover details of your life plan, on another level self-enquiry has to do with making a connection with your spiritual self at the node fragment level. There is also potential to establish a functioning connection with what exists far beyond your own spiritual self.

Awareness that this spiritual level exists is ultimately what drives the frustration people have with their lives. It sparks the restless feeling that something significant is missing. And it is responsible for the feeling that true satisfaction and peace lies somewhere else. Because, of course, it is in your natural home, the spiritual domain, that perfect peace and restful relaxation exist. Frustration and restlessness arise because you are aware, whether consciously or unconsciously, that perfect peace and restful relaxation have been temporarily lost to you throughout the duration of incarnation.

Thus the deep reward for spiritual work is the reminder that in spite of all appearances, there is indeed a point to life, and a safe place to return to at death. This process of venturing into unknown territory in any domain of existence, for the purpose of being strengthened by the trials one undergoes there, is universal. It also operates in degrees.

Some people seek extreme versions of venturing while incarnated and become the adventurers and explorers of the world. Others, less brave or driven, find quiet or secluded niches in a society to live out their lives in obscurity. All choices are equally valid, except to the extent they are driven by fear. This is because fear is the antithesis of love.

The conquering of fear is spiritual work, effectively producing love magnified by freedom of choice. It is to this end that we encourage the opening out of the personality, which is otherwise constrained by limiting early experiences. This involves therapeutic intervention, chosen according to what one finds fitting, and carried out in one's own long-term interest, with the aim of clarifying one's inner perception.

Self-enquiry, then, involves a multi-level effort. In terms of investigating one's psychology, it involves enquiring into the nature of the biological, socialised and essence selves, and appreciating how they interweave to produce the identity you manifest as you go about living in the world. In terms of investigating the details of your life plan, it involves delving into the nature of your essence self, because the higher human traits, abilities and talents that are present on that level directly reflect what your spiritual self wants you to engage with this time round. And in terms of investigating the nature of your own identity as a node fragment of Dao-consciousness, and investigating what is accessible beyond pure bodily sensory perception, it involves experimenting with the capabilities of your energetic and spiritual selves.

To practise self-enquiry, then, you have to learn how to disentangle yourself from the immediate moment, step back, and take a wider view of what is going on around you, to you, and within you. Human perception is narrow, whereas you need to adopt a wide view to perceive the interlinked factors, which overlap each other in often complex ways and that contribute to you being where you are, doing what you do, and appreciating what else you could be doing.

Pragmatism is the key to effectively practising self-enquiry. There are many forms of self-enquiry, which adopt different approaches to investigating human psychospiritual existence, and focus on different aspects of the five-layered self. It is up to you to choose a form of self-enquiry based on what you want to know and you find effective. In other words, you need to apply trial and error to find what works for you.

Self-enquiry begins with the acquisition of raw data. Self-observation provides the raw data. This is turn requires you to develop the ability to step back and observe yourself. One way of initiating this process is to keep a journal in which you regularly record what happens to you and how you react it, what you feel before, during and after, and what psychological, emotional and intellectual traits are active at the time. Dreams, meditation, prayer, along with anything else you do, also provide valuable observational data.

Next patterns need to be identified. There are many ways of doing so. All spiritual traditions, and a number of modern psychotherapy practices, provide a framework for categorising observational data and linking it into patterns of feeling, thinking and behaving. This is where pragmaticism applies. If you find a particular approach is illuminating and therefore useful, keep doing it. Alternatively, if an analytical framework becomes too limiting, move on to something else. The important issue is that you keep progressing. Progress occurs as a result of choosing a framework that works, committing to it, and carrying it out over an extended period of time. Consistency of practice over an extended period is crucial to making progress in any discipline.

Self-enquiry can be carried out in either a group or solo context. Again, it is a matter of what you find effective. One of the primary benefits of working in a group is that it involves setting a regular time when participants meet. Meeting once a week is traditional for metaphysical and self-study groups in recent times. This is arguably optimal, because it focuses participants' attention on their immediate issues on a regular basis. Establishing a regular meeting time clears a space in everyone's weekly schedule, providing an opportunity for deep engagement. However, do whatever is practical and possible. The important outcome is that meetings are held regularly and continued for a period of time.

If you are engaged in solo self-study, we recommend you similarly set aside a regular time to "meet up with yourself". This will ensure you maintain momentum. It is easy for a week or two to pass without any inner progress being made.

One notable advantage to group-centred enquiry is it provides a forum for sharing experience. When each brings to the group what they have observed and learned, knowledge is pooled, which is to everyone's benefit. When self-knowledge is shared by participants who respect one another's input, progress can be made relatively rapidly.

In the sciences anomalous data plays a crucial role in the furthering of knowledge. Anomalous data doesn't fit into current theories and models. If the data is accepted as valid, then it provides fresh information that challenges current knowledge. Anomalous data requires that accepted theories and models are either expanded, or rejected and new more inclusive theories and models developed.

Each person as they grow up are inculcated with the prevailing theories regarding life and death, and with current ideas regarding the nature of identity. For the vast majority of people, existence is entirely body-based, and any notions regarding spiritual nature and identity are considered far-fetched and largely unprovable. This applies even to the religious. They are taught metaphysical beliefs, but validation is provided by sacred texts and not by personal, empirical observation. In addition, everyone has a self-image that, as we have already discussed, is focused very narrowly around specific positive and negative traits.

Anomalous data in this context consists of perceptions that challenge whatever ideas regarding the nature of reality and how identity is constructed. Anomalous perceptions have the potential to explode your world view. Just as scientists seek anomalous data because it opens up new territories for investigation, so it is useful for spiritual investigators to actively pursue anomalous observations, because they open up new areas of experience and knowledge beyond the circumscribed boundaries of they have been taught.

Psychologically, anomalous behaviour refers to behaviour that seems to be out of character yet is repeated over a period of time. Often such behaviour occurs in periods of stress. It is at such times one is more likely to behave in a way that others consider out of character. It might be that an individual becomes angry to the extent that they explode. For some this might be a noisy explosion, even violent, whereas for others it might be a quiet explosion that they keep to themselves. Alternatively, the individual may become depressed, may feel everything is futile. Many people can get stuck in this emotion for extended periods. Other forms of anomalous behaviour include indifference, or self-pity, or responding in an abnormally childish way. Such behaviour is anomalous because it is not the individual's normal state.

When strong emotions generate anomalous behaviour, the occasion stands out and one cannot easily forget it. Unless, of course, one goes into a self-defensive mode and justifies behaving that way, or by blaming someone else for what happened, or ignores it, or denies that it occurred. Self-defensiveness is a manifestation of self-calming, a psychological self-stabilising behaviour that is widespread throughout the human population. When something challenging occurs, which upsets a person's usual inner state, self-calming automatically becomes active, its purpose being to return the individual's inner state to its former stabilised condition. In the case of anomalous behaviour that the individual finds disturbing, self-calming triggers defensive processes that end up consigning what was observed to the basement. The door is then firmly shut on it and the implications ignored. Of course, the problem is that when valid data is ignored no learning takes place, and therefore no progress is made.

Learning how to process anomalous data that is difficult to face up to, because it challenges what you feel and think, is a skill you have to develop in order to effectively practise self-enquiry. You have to be honest with yourself at all times. And to do this you need to learn to switch off the self-defensive and self-calming mechanisms that are naturally present in the human psyche.

Other forms of anomalous data comes in the form of dreams that are very strange, or a single dream that repeats night after night. In these cases, it could be that the spiritual self is attempting to pass on needed information to the individual's everyday mind. On the other hand, many dreams are manifestation of the autonomous nervous system's normal process of releasing bodily tensions. Accordingly, a repeated dream may not be from the spiritual self, but rather be an expression of unresolved tension. Yet this is still useful, because it provides anomalous perceptions, outside the usual everyday range, that are useful for seeing how your psychology functions and so offer data for enquiry.

Other anomalies are of a psychic nature. People occasionally see ghosts, have insights into what is troubling someone close to them, get glimpses of future events, or just have a feeling that they should do or not do something. Sometimes these involve communications from the spiritual level, at others they are spontaneous manifestations of abilities latent in the psyche that are not being explored but that are incorporated into the life plan and so have started "poking through" the fabric of everyday perceptions. Experimentation with these types of anomalous perceptions can be very interesting and fruitful. Again, it is an area where fear needs to be held in check, and honest appraisal of perceptions needs to take place, without falling back on easy explanations or externally determined judgements.

Then there are anomalous perceptions of a spiritual nature. Often these occur spontaneously, when one's awareness opens out from its normally constricted state. Spiritual practices can foster these types of anomalous perceptions on a regular basis. They can also lead to encounters with non-embodied identities. As with anomalous psychic experiences, fear needs to be held in check and assumptions regarding what is being perceived avoided.

Having already spent some time discussing psychological factors, we will now examine anomalous psychic and spiritual perceptions in a little more detail, beginning with the nature and function of the aura.

CHAPTER 7

THE FUNCTION OF THE AURA

THE AURA IS SIGNIFICANT BECAUSE IT PROVIDES a communication channel between the spiritual and physical domains. It provides a natural conduit by which anomalous perceptions arrive. But it also has other functions that first require comment.

We identify the aura with Bohm's implicate order as explained in the first chapter. In the sense we are discussing here, the aura is a structure within the implicate order. This means the aura is electrospiritual in nature. One of its key functions is to contribute to the body's form and natural qualities. Earlier, we asserted that the implicate order provides blueprints that shape the general patterns present in the electrospiritual into species' particular bodily patterns. The aura facilitates this process.

The aura's emergence from the implicate order begins when the sperm first fertilises the egg and the two fuse to form a zygote. The aura can be thought of as a local accretion of the electrospiritual that condenses around the foetus. This is not unique to human bodies. Every body of every single living creature, including plants, insects and animals, possesses an aura.

For each species, specific patterning relevant to a creature of that species manifests out of the implicate order. The aura's function is to channel that information to the growing body's cells. In effect, the aura is the means by which the plastic undifferentiated cells are imprinted with instructions from the implicate order regarding their particular location on the endoskeleton, and fit themselves to their appointed biological function—bone, flesh, brain tissue, and so on.

Obviously, the genes also contain information relevant to the development and formation of a specific body. These relate to details: hair colour, eye colour, propensities to certain abilities and illnesses. In contrast, the aura transfers broad species information to the foetus. Therefore the aura may be described as performing a dynamic organising function for the construction and maintenance of an organic body. At the end of a body's lifetime, when the need to maintain that physical organism ceases, the aura returns to that from which it came, the electrospiritual.

Why are we making so much of the electrospiritual in relation to foetal developmental? And how is it that the implicate order is so significant in particular to humanity? The answer is that the human aura was shaped in the distant past by Dao-identities whose aim was to bring the human form to its optimal condition so node fragments could co-associate with them and partake of the most complex experiences possible within the human domain. We note that this type of auric level modification no longer takes place. The association of the growing aura with the growing body is now an automatic process. The aura is imprinted by the implicate order, and the aura automatically imprints the growing cells.

As we have also noted, before incarnation a node fragment selects certain specific character traits and abilities. Some of these will likely already exist in the genetic makeup of the selected foetus, being factors that contributed to that body being a suitable option. But other traits, not supported genetically, are also transported to the new identity. This process could be termed fine-tuning, in the sense that it involves subtle information regarding character traits, predilections, what human beings call natural abilities, and so on, being embedded in the nascent identity. This information is embedded at the auric level.

So the aura contains two levels of information. One is impersonal, being general information regarding the shaping of the cells into a human body. This information comes from the implicate order and is automatically encoded into the aura. The second level is personal. This is information that the node fragment deliberately imprints into the aura. It consists of traits and abilities selected to facilitate specific kinds of life experiences.

This doesn't mean that a node fragment is able to transform its newly selected sub-identity into a great sports person or musician. It cannot transfer traits that it has not already developed in prior lives and that are not present in the node fragment's accumulated human identity. What it can and does transfer are traits it has developed through its own efforts. It imprints these into its new body's aura and so influences the development of its new identity. This occurs because a node fragment recognises that certain qualities will optimise the physical and psychological nature of its new incarnation. All this occurs at a level of which any currently embodied individual is unconscious.

When anyone instigates a process of self-enquiry in order to learn what qualities they have brought into this life, and why, they access the relevant information via their aura.

Besides patterning physical cells, the aura also acts as a natural channel of communication between a node or node fragment of Dao-consciousness and the embodied human being. We have just discussed how, when the foetus is growing, the aura transmits information from the implicate order to the growing cells that contains instructions regarding how and where to grow. In this sense, the aura acts passively, as a conduit.

This passive quality also makes it ideal for deliberate and consciously directed communications from the Dao-identity level. And because the aura is a passive conduit, communication easily goes both ways. This means that when an individual functioning at the level of its ordinary mind wishes to communicate with its Dao-identity, the aura facilitates the flow of that communication.

This function of the aura is seldom appreciated. Often when people experience communication from a node of Dao-consciousness they think something divine has touched them, or that they have been specially chosen. In fact, it is the aura that makes non-ordinary communication possible, which in turn provides anomalous observational data.

On the other hand, such communication can also generate fear. When auric level communication is first initiated many people experience a great shock. They are confused by the signal. They may deny anything has happened at all. Then, if it is accepted that something out of the ordinary has happened, there is much uncertainty as to where the contact comes from. For those with a strong religious upbringing, there may be fear that such activity is not allowed, or even that the demonic is involved. For those with atheistic or materialistic conditioning, there may be no willingness to acknowledge that something they cannot explain has occurred. In each case, the contact is likely to be declined.

Psychologically, the unease can be said to result from unfamiliarity, given that unfamiliarity often leads to fear, denial and repudiation. Such responses may be likened to the unease people feel when others around them are speaking in a foreign language. They wonder what is being said, and especially whether they are being talked about negatively. If their unease is acute, they may begin to feel threatened. Clearly, their discomfort arises out of their lack of knowledge of what is being said, and particularly from their ignorance regarding the intentions of those speaking in the unfamiliar tongue. The same applies to auric communications. People reject, deny and fear them out of uncertainty regarding what is behind them, and especially because they are ignorant of the level on which such communication occurs.

Fear is key. Fear of the stranger can be seen to be behind the discomfort people feel among those speaking in a foreign tongue. The same fear, bolstered by fear of the unknown, applies when people are faced with communications coming from non-embodied identities who use auric channels to disperse information.

The way to combat fear is to foster familiarity. Familiarity with the way that auric level communication takes place, and the reasons why it is occurring, generates a feeling of familiarity and consequently of comfort. Practice fosters both.

A number of different kinds of spiritual-level communications are possible due to the aura's capacity to function as a passive conduit that facilitates the flow of information. Auric communications initiated by one's own spiritual self are the most common and the most frequent. For many people they are also the most frequently declined. Simple uncertainty is the usual cause, allied to immersion in daily life that prevents the recipient from paying proper attention. Life activities can be so demanding that the recipient is too tired or unfocused to absorb the fact that communication has been initiated. Most such communications from the spiritual self to the everyday mind are to do with fulfilling life plans, as already discussed.

Another reasonably common type of communication emanates from spiritual level friends or guides. Such communications are often intended to provide a psychological jolt, or what is colloquially called a wake-up call. This occurs when the individual is confused, has departed from the life plan, has descended into self-damaging behaviour, or has a significant life event on the horizon and the individual's own spiritual self is unable to bring to the individual's attention what is intended. There are many ways such a communication may occur: as a dream, as a shout in one's head, in the form of a vision, drawing attention to a book or image, a powerful feeling ... the list goes on. What they have in common is that the communication is initiated out of goodwill, and the aura is affected in some way in order to draw the individual's attention to the act of communication.

Spiritual identities initiate this kind of communication in order to help in some way. Often it is in response to a call for assistance. Typical circumstances include providing information, especially offering a different perspective, when a big decision is required. Another common circumstance is giving an individual who is feeling unhappy, depressed, lost, or uncertain an auric level hug, so to speak, so they feel supported and not alone. Often support is given to help an individual in relation to completing specific tasks related to the life plan.

A less common form of auric level communication involves the channelling of information such as is being provided in this text. In the case of the two individuals involved in helping us share our knowledge with embodied humanity, they have established channels of communication via their auras. A common way to make contact with such persons is to exert pressure on their aura. This pressure may be experienced as chilliness, warmth, as tingling, as agapé, or in the form of many other physical or emotional symptoms. This process provides an important piece of information about the relationship between the aura and the biological self. Basically, the aura, as part of the energetic self, can send impressions directly to the body that are experienced via the body's senses but that initially bypass the everyday awareness. The reason this occurs is that everyday awareness is usually filled with the activities of daily existence. So when information emanating from the spiritual domain becomes available, the everyday awareness is too occupied to notice. An example will make clear what we mean.

Imagine a house in which the kettle is boiling and whistling. (It is an old-style stovetop kettle that doesn't switch itself off.) However, Dad is in the garage working on the motor mower, Mum is vacuuming, and the children are playing a noisy game. So no one hears the kettle whistling. And even if Mum turns off the vacuum cleaner the noise from the kids and the engine in the garage may still be too loud for her to hear the kettle. But on this occasion something in her attention "clicks" and she remembers she left the kettle on, so she goes to the kitchen and turns it off. After Mum turned off the vacuum cleaner, and before she had time to occupy herself with her next task, there was a gap in her attention. In that split second the information came that she needed to turn off the kettle. She responded at the level of her biological self before her everyday awareness engaged with the information and wondered where it came from. Thus her biological self had a more direct route to the information in her memory than did her everyday awareness.

This works on the passive level. But it also works on the active level, when spiritual identities external to the incarnated individual seek to make communication. They make their presence felt by "pressing" on the aura. That pressure manifests on the level of the biological self and the visited individual's everyday awareness is then drawn to what is occurring on the spiritual level. In this sense, such "pressing" may be likened to a guest arriving at a house and pressing the doorbell to announce his or her arrival. Exactly the same announcement process occurs when the aura is "pressed".

Another of the aura's functions is to facilitate the flow of spiritual-level energy during the process of spiritual healing. We are speaking here specifically of the situation in which a healer opens up his or her awareness to input from the spiritual level and intentionally directs that input to a person who seeks to be healed.

For such a flow of energy to occur, several pre-conditions need to be in place. The healer has to commit to working with non-embodied identities in order to heal others, and has to establish an agreement regarding how they will work together. The healer then needs to learn about energy flows, in particular how to utilise spiritual energy for healing and to avoid drawing on their own energy. The healer also needs to learn how to seal themselves off from the sick people they are helping, so they will not be contaminated by the recipient's energy. The prime energy involved in healing is the force of love, so to be an effective healer the nature of love must be understood. All this is involved in the healer learning, through a trial-and-error experimental process, how to facilitate the flow of spiritual level energy through them, via their aura, to the embodied person they are helping.

Our final example here consists of those who communicate on behalf of others, who possess what are popularly called psychic abilities. They have established a means of using their aura to communicate at an energetic and spiritual level with non-embodied identities who wish to communicate with human beings. This category largely consists of mediums who pass on messages between the living and the so-called dead. We say "so-called dead" because in fact they are all living, just in different domains.

Because each person's aura has been associated with their body since their body was conceived, once a communication channel has been opened between the human personality and the spiritual identity information flows easily and naturally. It is just a matter of opening up one's mind to the possibility, stilling thoughts and emotions, and sustaining the intention.

The process of communicating with the spiritual realm involves an embodied everyday mind using the aura's capacities to receive information directly from the spiritual domain. However, it is also possible for an embodied individual to communicate directly with those in the spiritual domain by projecting their awareness into that domain. We refer to the practice of meditation.

CHAPTER 8

THE ART OF MEDITATION

WHEN MEDITATORS CLOSE THEIR EYES, still their minds, and "look out" into the mental void, they project their awareness into spiritual space. In ancient times meditators were known as shamans, and the act of meditation, during which an embodied individual's awareness explored spiritual space, was called shamanic flight.

To ground this discussion in human experience, and to show that what we are discussing is by no means new but has been practised throughout human history, to the traditional notions of shamanic flight into shamanic space we add the model of agapéic space. For our purpose, it may said that agapéic space refers to spiritual space in its theoretical construct, while shamanic space refers to the explorer's subjective experience of spiritual space.

In using the term shamanic space you will note that we have chosen the most ancient, and therefore non-religious, terminology. It is not the realm of God, heaven or hell, for they are all constructs. Yet, those ancient notions, which are derived from experience, may be subsumed into the notion of shamanic space. We simply seek to make this an a-religious description.

It is difficult, if not impossible, to find ancient descriptions of shamanic space that agree well with one another. This is because so much historical material is lost, and much of what remains has been subjectively reconstructed long after the fact. It is into this confusion that we are attempting to introduce a little order so any investigator into what is conventionally termed spiritual experience may feel an increased level of confidence about what they are exploring.

Regarding the phenomenon of shamanic flight, the shamanic world view commonly describes a lower world, a middle world and an upper world, and places different denizens in each. Yet traditional descriptions of shamanic flight do not clearly distinguish between the different modes of exploration. This is due to the relatively recent establishment of psychological language that is now used to describe personality and its experiences.

Descriptive language is so powerful that it shapes an individual's self-identity and perceptions, with each being interpreted through whichever language is possessed. People in different cultures use different concepts, interpretative frameworks and language. As a result, interpretations of spiritual experiences are culturally-bound, and as a consequence frequently differ from one another.

It is very difficult to provide culture-free interpretive descriptions of spiritual experience—in fact, it is an ideal that will likely never be achieved. Nevertheless, adopting a single descriptive language enables experiences to be compared for similarity and difference, even if certainty that compared experiences are exactly the same is unlikely. With that clarified we now continue our exposition of shamanic flight within shamanic space.

Meditation requires that individuals begin in a state of being perceptually open. This is achieved by activating the brow chakra and extending perception via that energetic node. When individuals tune into the agapéic level by directing their focused attention via the brow chakra, they become aware of a very large perceptual space in which they themselves are located, as if in the middle of the sky.

One way for those who are inexperienced in this type of perception to accustom themselves to the experience of being in shamanic space is to imagine their body boarding a plane, the plane taking off, flying high into the sky, and entering a cloudy space. There imagine, within that cloudy space, the plane stopping but not falling. Then picture the plane dissolving, or disappearing, or falling out of the mental picture, leaving the individual suspended and stationary, experiencing only that cloudy space. That is the perceptual analogue of what is experienced when individuals directly perceive themselves being within agapéic space. There is nothing around them. They are present but going nowhere, and they cannot see anything. In fact, commonly there is no perceptual input via the brow chakra.

Initially, there appears to be no light in that cloudy space. Therefore the fog seems to be black, or at least very dull, and no matter how much the meditator tries little is seen. However, if the meditator spends enough time in that initial condition, being content to direct their attention via the appropriate perceptual channel, with practice that door of perception can be cleaned, as it were, and eventually the meditator becomes aware of being suspended in agapéic space observing the variety of movements of individuals in proximity. This is because others present in that domain eventually recognise the meditator as being an object of interest, with whom communication may potentially be made.

A variety of fears have to be overcome in order to participate in this practice. We begin by detailing some of the most common fears.

The first fundamental fear is that of survival. The body's concern regarding the physical extinction of both it and of the everyday mind that identifies with the body can trigger animal survival mechanisms. However, explanations may help individuals transcend that fear.

As discussed earlier, all individuals encounter their first significant fears during childhood. Perhaps, on one occasion, they were unceremoniously tossed by others while playing and were shocked to experience unexpected flight. To counter this fear of being untethered and flying in an uncontrolled manner, recognise that in meditation there are two means by which the mind can move. First, the mind can move of its own accord. Second, it can be taken.

The prime responsibility for taking the everyday mind resides at the level of the spiritual self. Given the spiritual self is concerned for and actively nurtures the individual, no risk need be associated with the idea that the loving spiritual self may take the individual's attention out of the body. Of course, understanding that at a conscious level does not necessarily cancel all self-concern or eliminate active resistance.

Visualising being submerged in water, yet floating in it, provides another metaphor for the experience of being immersed in the sea of shamanic space. It could be imagined that one is immersed at a depth of thousands of kilometres and so is unable to escape. Yet reflection reveals that being immersed in shamanic space is actually a normal and safe experience, providing as natural an environment for humanity as water provides for fish, given human beings always exist in spiritual space, whether embodied or not. Recognising that spiritual space is a field of existence beyond time, and therefore that identity cannot cease to exist, potentially reduces residual body-based fears regarding survival.

It may also be imagined that tides can carry one to this place or that, or that one's own initiative can do likewise. One can visit points of attraction or migrate there permanently. Some places are darker and others lighter, some higher, some deeper, some here, others there. But all exist within that intrinsically spiritual environment. In this way we associate the metaphor of deep water with the traditional metaphor of shamanic flight, in which one proceeds from one's body location in the middle world and goes to the upper or lower worlds.

The reality is that embodied identity encompasses both the physical and spiritual. Yet when the spiritual self approaches the everyday mind, that mind normally projects a sense of otherness onto it. This leads to barriers being raised and the feeling that such encounters are potentially threatening—threatening, that is, to the body-mind identity. This simple perception, that one is being approached by a foreign "other," is the basis of much fear. It is necessary to allay such fear before what could be thought of as comfortable cohabitation between the lower everyday and higher spiritual aspects of identity become possible.

With an appreciation of the nature of spiritual space, and that the everyday self need not be threatened by what it perceives there, fear may be set aside and meditators can then comfortably park their body, as it were, and go flying.

In contrast to the exaggerated descriptions generated throughout history, some of which has been converted into coercive literature that extols dangers to be avoided and joys that may legitimately be sought, the experience of spiritual level journeying is rather mundane—even though it involves encountering seldom accessed dimensions and utilising seldom tapped modes for acquiring information.

However, before embarking on such journeys, it is helpful to be forewarned that tricks and traps exist in the lower mind, and that there are methods of defence, and pitfalls to avoid, which will help the lower mind maintain its equanimity and integrity. Distinction also needs to be made between this level of journeying and experiences undergone when the lower mind is transcended altogether, an occurrence discussed in literature that confuses the two levels of journeying and the different kinds of experiences they generate.

Accordingly, we attempt to disentangle these categories by assigning two distinct territories existing within the lower everyday mind: the imagination and related internal defense systems, and the higher mind manifested by the spiritual self and located in agapéic space.

By way of an example, we draw attention to what occurred when this individual [*Editor: Peter*] sought to simplify his inner processing by confronting the structured layers of fear that existed within his lower everyday mind. That mind mounted a defence by generating the image of a large wild tiger leaping towards the point of attention. A more timid personality would have quailed at that point. This individual's personality was able to see that event as the ephemeral defence it was, and proceeded regardless towards simplifying and clarifying its mind. Accordingly, when the objective is to address the everyday mind itself, expect a defensive reaction. Resist whatever imagery results and exercise goodwill, and the task is more likely to be successful.

One way of distinguishing between self-directed intention at the level of the everyday mind and intervention and phenomena occurring at the level of the spiritual self's higher mind, is that higher mind phenomena is usually identified as not-me. That is, what becomes apparent to the attention is more likely to be interpreted as involving others who are independent of oneself, having distinct characteristics and experience. Once the distinction has been successfully made between the different levels to which attention may be directed, exploration of these dimensions, spaces and identities follows. One can then proceed even further and direct the point of attention beyond both the lower mind and the higher mind and out into the Void, where identity is absent and one can attain the experience of vast emptiness.

It may not initially appear to be so, but the individual has choice regarding what is experienced. With repeated experience one develops the capacity to expand the range of one's exploration and go into zones presenting different illuminative experiences characterised by colours. There is the ordinary proximate dark zone, which is actually a clear spacious darkness of minimal illumination that progressively lightens. With the substantial movement of the point of perception the colour changes to white, then gold. That gold perception is accompanied by a feeling of encountering a loving nature, because gold contains universally attested compassionate love. And so, within these general parameters, one has a basic map of zones of experience that correlate with the lower mind, the higher mind and the Void.

This brief general outline comprises a map of the territory. It is sufficient to enable a novice investigator to place the content of their experience into one of the three categories. The misperceptions and misattributions so easily made in these classes of perception mean it is best to be tentative in assigning a category to what one experiences. As experience accumulates one will develop greater certainty that one has accurately identified where one has got to during these types of interior investigations.

CHAPTER 9

AGAPÉIC SPACE AND ITS CORRELATES

THE NEXT ISSUE WE WISH TO DISCUSS is the relationship between meditators' capacity to use their point of attention to project themselves to distant places and states, and their ability to process that information using intelligible models to minimise the possibility of the information being misinterpreted.

Perception of interior phenomena is impacted by a layer of preconscious processing. Preconscious processing refers to concepts, assumptions and models embedded in the meditator's mind. In most cases they are learned, being placed there during childhood through education and enculturation. These pre-existing concepts, assumptions and models are then used to interpret, and as a result colour, an individual's perceptions. An example of preconsciousness processing is the common assumption that when a light is experienced within it means a favourite saint, avatar or god has been encountered. It is rare for that actually to be the case. However, people need to create meaning for their experiences and therefore use models they are familiar with, whether they be mythological, psychological, spiritual or random selections.

Preconscious processing can be so subtle that the everyday mind is unaware it is occurring. This then reduces the mind's capacity to affect it. However, knowing that preconscious processing occurs means a degree of intelligent choice can be made at the post-conscious level, when a meditator reflects on perceptions. That being the case, a consciously selected framework can aid in the task of questioning interior experiences and interpreting them into a new narrative. We propose that new framework needs to be empirical in nature. By this we mean that when a meditator says, "God spoke to me through a cloud," that experience could more accurately be interpreted by saying, "There was an impression of light, an impression of cloudiness, and I heard a voice but I didn't see anybody." In this way an empirical description of an inner perception replaces a theological interpretation that overlays and confuses the content of that perception.

Throughout history, individual's interpretations of their experiences have been reported as being the actual experience. That erroneous reportage has then been taken up by others, who were predisposed to interpret their own experiences using the same terms. As a result, patterns of expectation and interpretation have grown within all cultures, each with characteristics that differentiate one culture's interpretative framework from others. This explains the differences in visual imagery applied in different cultures and religions. In most instances such imagery derives from an accumulated history that has been distributed through a community, from which individuals acquired models they subsequently used at the preconscious level to interpret whatever interior phenomena they observed.

In proposing empirical descriptions be favoured over culturally generated models, we are attempting to track, refresh and modernise the models used for interpretation, and especially to dissociate interpretations from traditional religious descriptions. It is not that religious descriptions are necessarily invalid. They are certainly not, given these events happen and always have throughout the history of shamanistic perception. However, most people have little experience of these kinds of phenomena, with those who find themselves receiving such specialist observations being regarded as naturally gifted and sensitive. We seek to harmonise the terms individuals use when perceiving and labelling meditative perceptions, providing a fresh, coherently organised framework within which all can recognise observed phenomena as genuine, then, wherever possible, categorise and locate them accurately, without misattribution.

We understand this goal has an innate weakness, as mythological and religious metaphors are so ingrained into all cultures that they will only be eradicated with extraordinary effort. In some cultures, science's domination of religion has moved them some way towards achieving this. But to date it has proved impossible to eliminate all such influences, even in the most advanced cultures.

Given this, we offer our best attempt to provide a high quality interpretative framework for the modern community with the aim of re-setting the clock on the patterns of observation derived from first-hand experience. Accordingly, our interpretative framework is grounded in what meditators directly perceive while exploring agapéic space.

When meditating the point of attention is normally felt to exist within the braincase. The mind uses the body's physical senses, and primarily the visual sense, to orient itself. When the point of attention moves out of the braincase and into the spiritual realm, meditators equally need a way to orient themselves. We propose orientation be grounded in the meditator's experience while seated in a meditating posture. This means that just as the meditator's mind ordinarily interprets local space in terms of the three axes of left and right, forwards and back, up and down, so the detached mind exploring agapéic space can orient itself using the familiar axes of left and right, forward and back, up and down. What complicates the use of these three axes in agapéic space is that the freedom individuals have to move their awareness within the spiritual domain is in proportion to the development of their inner capacities.

The word *agapéic* derives from the ancient Greek, *agapé*, meaning fraternal love. The term was later adopted by Christians to refer to spiritual love. We call the spiritual domain agapéic space because human beings are able to consciously access it via love. As a result, the three orienting axes of left-right, forward-back, above-below may be characterised in developmental terms as willingness of bequest agapé, agapé frequency and hierarchy.

Agapé is intrinsic to Dao-consciousness, and to the node fragments that co-associate with the human species. Agapé is characterised by various qualities, including loving regard, a desire to help, wishing the best for any other, self-concentration into goodness, willingness to love all others, and possessing an optimistic and positive outlook. Other attributes include a desire to nurture others, being willing to focus on any other individual so that other-interest predominates over self-interest, a capacity to hold a group in special relationship, willingness to see beyond the superficial personality to the core of another person's nature, recognising that any other is intrinsically and at core identical to oneself, and claiming commonality with all others, human and non-human.

So the first developmental axis, willingness to bequest agapé, directly reflects the meditator's capacity to love. We note that this is a quality that needs to be fostered in one's relations to others in the world, and the more it is fostered, and the more personality traits that diminish the sharing of agapé are eliminated, the greater inner capacity meditators will possess when they enter agapéic space.

Willingness to act from agapé gives rise to hierarchy. Hierarchy constitutes the sum and product of loving acts enacted throughout a life.

The term *hierarchy* in the sense we are using it here is not the same as the social hierarchy promoted in human cultures. Socially, hierarchy denotes one person being above another, whether by virtue of occupation, wealth, heredity or celebrity. Instead, we use the term hierarchy in the sense of progression, a developmental progression. Agapéic hierarchy doesn't result in one person being inherently more spiritual than another. It merely denotes their greater capacity to love.

For the sake of discussion, we designate hierarchy as comprising one hundred steps. Each step may be subdivided in order to accommodate an increase in hierarchy due to development over a nominal thousand lifetimes. Hierarchy is usually added up at the end of a life, which is the reason movement in hierarchy is most often assigned at the end of an incarnation—although some rare individuals may carry out loving acts of such power that their hierarchy is raised while still embodied.

For agapé frequency we propose a scale that extends from zero to sixty-five thousand. These are notional numbers only, from which we have generated a model to indicate the nature of human awareness within the extension of all agapéic space. Within the sixty-five thousand range, human beings exist from 25,000 to 35,000. This directly reflects their development over their one thousand or so incarnations. Naive, inexperienced node fragments of Dao-consciousness begin their incarnational cycle at 25,000 on the agapé frequency scale, and that cycle ends when they achieve an agapé frequency of 35,000. So this is another developmental scale. It also makes apparent that human capacities occupy a very small range within the totality of what is possible. Moreover, development continues after the incarnational cycle is completed and the fragments of a node reintegrate. We, as reintegrated node fragments, each experienced a full human incarnational cycle, and now occupy a position of 42,000 of the agapé frequency scale.

These three axes extend across the shoulder blades (willingness to bequest agapé), forwards from the hara dantian (agapé frequency), and upwards from the spine (hierarchy). We note that agapéic space, with its three axes, is a *model* of spiritual space. It is a conceptual framework for explaining certain arcane aspects of Dao-identity as it exists in its natural non-embodied state within spiritual space. Just as a map is not identical to the terrain, so agapéic space is not identical to spiritual space. It is a description of certain aspects of spiritual space.

As a model, agapéic space has two uses. The first we will describe in what immediately follows. The second, which draws on the agapé frequency scale, we will discuss in the next chapter.

The three axes proposed in the model of agapéic space enable meditators to appreciate the intrinsic nature of spiritual identities they encounter while exploring shamanic space. As a volume, agapéic space is larger than any person can imagine. It is also compartmentalised into zones which are occupied by every species in all the universes. No species' zone of existence encroaches on any other species' zone. Pragmatically, from a meditator's perspective, what this means is that spiritual identities, being nodes and node fragments of Dao-consciousness, may be encountered during meditation—or, we add, in visions or dreams. These identities possess their own agapé frequency, and gift agapé to various degrees, or not at all. Such is the diversity of identity at the node level, reflecting different aspects of the Dao, which come to the fore within them in a wide range of propensities.

This means that the node identities meditators meet are not necessarily similar to themselves. They have different capacities and intents. As a result, some encounters meditators have will be pleasant, informative or invigorating, others less so. We do not say this to scare anyone, or to suggest agapéic space is a dangerous place to explore. Far from it. We are just making plain that the beings human meditators meet in agapéic space are not necessarily human-like.

The point of the three axes of agapéic space is that they offer a way to identify the intrinsic nature of those who are encountered. Those who come into the meditator's perceptual field from the upper right possess greater hierarchy and higher agapé frequency, and so are invariably trustworthy. Those who come from the lower left quadrant possess lesser hierarchy and agapé frequency and so may or may not be trustworthy. Those who move with apparent freedom in the generally

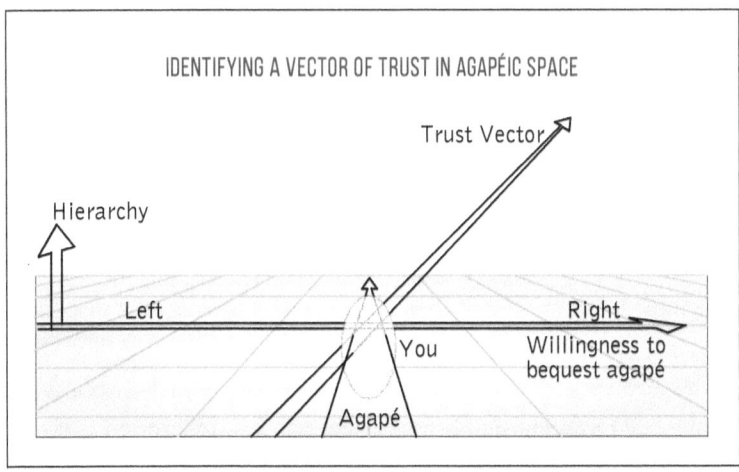

forwards direction may be doing so as a test of the individual's perceptual capacity to track their movements. And those who rise up from the lower rear left quadrant are generally to be greeted with loving goodwill while maintaining alert awareness with regard to their energetic signature.

It must be noted that this model has been generated to help meditating human beings appreciate differences in the natures of those they interact with in agapéic space. So it is entirely a human-level concept. Once an individual's body dies, and their awareness is no longer tied to the physical domain, these references become obsolete and other frames of reference apply.

We finish this explanation by noting that meditators may perceive differently when they explore agapéic space. Fear is one reason. When something new is experienced fear can colour the perception, leading to an encountered identity seeming to be a threat when it is not. This involves projecting base emotions, which shade the encountered identity with intentions it doesn't have. Another reason is that preconscious processing causes meditators to confuse their interpretation of their perception with what they are actually perceiving, as discussed earlier.

A third reason is due to natural and often unavoidable disturbances that occur in the correlation between an individual's bodily structure and shamanic space. This may happen because of tiredness, or because the body is reacting negatively to substances ingested to induce a experience of agapéic space in the first place. Whatever the cause, it leads to individuals losing their orientation, with the result that they don't know where they are in terms of the three agapéic axes. This leads to an incorrect attribution of what they experience. For example, in ancient shamanic terms, they designate an identity as situated in the upper world when it was actually situated in the lower world. In terms of the agapéic axes, they might wrongly designate an identity as being higher and to the right of them when that identity is actually on the same level but to the left.

Nevertheless, we maintain that the correlation between bodily structure and agapéic space holds, because this description is coming through an embodied individual and it is designed to speak to other embodied individuals. Because it is specifically linked to the human sense of orientation in space, the experiences of spiritual space had by spiritual identities co-associating with non-human species are not relevant to the description we are offering here. It is highly likely they would not find this model illuminating or useful.

CHAPTER 10

THE AGAPÉ FREQUENCY SCALE

WE OBSERVED IN CHAPTER THREE THAT it is the Dao's nature to manifest nodes. Nodes are individual agglomerations of consciousness that, when cast from the Dao, have various sizes. Some larger nodes fragment, most do not. However, almost all nodes associate with physical domains or species in some way. This means that nodes and node fragments are active throughout the multiverse.

On this planet there are many different nodes associating not just with biological species, but also with every ecosystem and environmental niche. Nodes possess different levels of complexity. When they seek to co-associate with a physical species, nodes select a physical species that has a level of bio-complexity that matches and is appropriate to their Dao-complexity. This being the case, it is useful to identify the range of those identities as they exist on this planet.

This identification of nodes offers a second use of the agapé frequency scale. The purpose of this scale is to create a conceptual framework human investigators may use to conceptually order those identities relative to each other. The agapé frequency scale extends from 0 to 65,000. As with hierarchy, these are arbitrary numbers, generated for illustrative purposes only. As with the concept of hierarchy, no sense of higher or lower, greater or smaller, is intended by this scale. This scale may be likened to identifying the size difference between an oak tree and a daisy. A single oak tree is larger, and so has the capacity to absorb more sunlight, rain and minerals than can a single daisy, yet one is not more important or greater than the other. Similarly, one node may be larger and so possess more Dao qualities than another node, but that doesn't make it greater or more significant. Each is what it is.

Before beginning, we need to clarify that this scale as we are about to apply it refers not to biological species but to the spiritual identities who associate with various biological species. Accordingly:

For those nodes that associate with plants we accord the range of 5 to 10. [*The numbers used here have a multiplier of 1,000.*]

Nodes that co-exist with group or hive minds, for instance, vertebrates such as aquatic species, avian species, and nests of social insects such as bees and ants, exist in the range of between 10 and 15 on the agapéic frequency scale. We also place invertebrates such as worms, and those with exoskeletons, such as crustacea, within this range of nodes.

Spiritual identities that co-associate with warm-blooded mammals we place on a scale of 15 to 20.

Elementals, those non-embodied identities that oversee vegetation and environmental niches, we situate from 20 to 25.

Spiritual identities that co-associate with human bodies extend across a range from 25 to 35. By this we mean that when an inexpe-

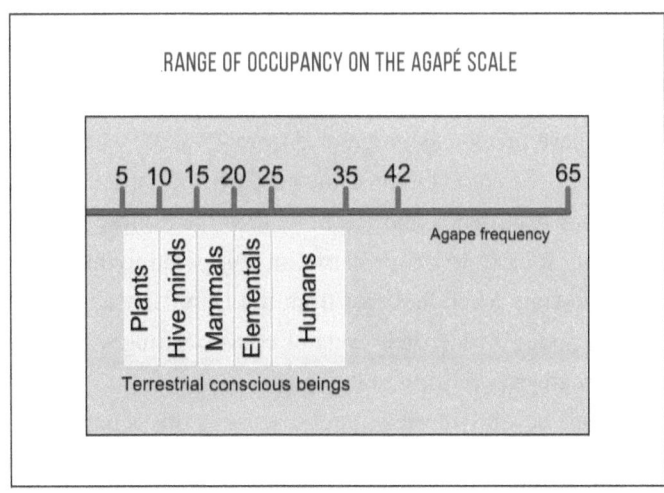

rienced identity first associates with a human body it is at 25 on the agapéic frequency scale. As it matures, its agapéic frequency increases. When it reaches 35 on this scale, and when it also achieves maturity on the other two axes (agapéic hierarchy and willingness to bequest agapé), it graduates from the incarnation cycle and rejoins other fragments at the same developmental stage who are beginning the process of reintegrating the fragments into a reunified whole.

The spiritual identities that co-associate with the cetacean family of whales, dolphins and porpoises extend from 20 to 30. So there is an overlap with humanity that makes it possible for node fragments of this complexity to migrate to human experience, although not many choose to do so due to the demands the human body and human interactions place on them. It is more common for identities co-associating with the human species to spend time among the cetaceans. Although even this is infrequent, as most of those rare individuals who seek an alternative body form and social experience select species from other planets.

Note that, like the human, there is a frequency range for the identities that associate with various forms of physical life. As with the human, this range indicates a developmental scale. That is, identities begin co-association at the lowest point in the range of agapéic frequency, and as they mature they progress to the highest point within that range. They then graduate from embodiment.

We situate ourselves, as a reintegrated node of Dao-consciousness, at 42 on this scale. Given that we were once embodied node fragments, and that we have previously progressed from 25 to 35, what this model implies is that nodes of Dao-consciousness are not limited in their evolution. As they experience and learn, so they themselves expand their capacity, in most cases to a radical extent, compared to their naive and inexperienced state when first cast from the Dao. Hence this scale applies only to nodes during their cycle of co-associating with a physical species. They all grow beyond that.

The range beyond 42 we will not discuss as this is not our brief in relation to this transmission.

An over-riding human assumption that we wish to counter with this model is the assumption that human beings are superior to other life forms and that human consciousness is the pinnacle of creation. Religions have historically promulgated both myths, telling stories of how God gave human beings dominion over all other creatures, and that only human beings have a soul. Equally, the scientific perspective has been that the universe is dead, that human beings are a fluke of nature, and that there are no other conscious beings in the existence. These are all fantasies. The universe is teeming with life, nodes of Dao-consciousness exist everywhere, the species Homo sapiens sapiens is far from unique, and human awareness is far from the pinnacle. As a broad generalisation, we would situate human node fragments in the middle of the entire range of nodes of Dao-consciousness. This is discarding outliers, some of which are tiny and others are massive and on a scale beyond human conception.

In order to give an example of life for other nodes co-associating with biological species on this planet, we will discuss dogs. Most nodes that associate with species do not fragment. Nor do they incarnate within an individual of a chosen species. The fact is that most nodes associate with a number of individual animals within a species. How many depends on both the complexity of the node and the complexity of the animal. However, nodes that utilise the canine experience do fragment like those that associate with human beings, just into a smaller number. On average they number between 60 and 150 fragments. This is indicative only, as some entities fragment to a greater extent and others to a lesser. But these numbers provide an average range.

The nature of these fragments differ from the fragments that occupy human beings. However, as many pet owners will attest, love, devotion, tenderness, and care are all qualities present in those spiritual identities that associate with the various canine species. It is also clear that at times instinctive, self-protective animal traits manifest in dogs, so rage, violence, sexual urges, an ability to identify friends and threats via smell, and so on, naturally dominate the awareness of individual dogs. Socialisation also impacts on the canine awareness, so trust and distrust, encouraged helpfulness to others and encouraged greed, manifest in the behaviour of individual canine animals.

All this means that the dog awareness is layered similarly to how a human awareness is layered, with animal and instinctive urges, socialised behaviours, and an over-arching spiritual identity, all manifesting in any individual doggy identity. Like human beings, the spiritual fragments that associate with dogs do so repeatedly as they seek to experience, learn and evolve. The exact numbers of associations differ between species, with some doing so dozens, others hundreds, of times.

The reason for these differences is that canine minds differ in complexity between species. Some doggy brains are rich and complex, especially emotionally, while others are less so. Accordingly, different complexities of node fragments associate with different canine species. More complex fragments associate with species possessing more complex brains, while the less complex associate with species possessing less complex neurology. For the same reason more complex fragmented identities incarnate more times because they are able to absorb a wider variety of experiences and at a greater depth. So they obtain more from their successive incarnations. In this way a spiritual fragment enriches itself and has more to bring back and contribute to its overall identity when the fragments recombine.

We would note that within the canine world a very wide range of experiences are available. The life of the domesticated canine is the most common, but domestication extends from a working dog on a farm to that of a pampered pooch that spends its life entirely inside. There is a life in the wild as a wolf, or a different kind of untamed life in the wilds of urban streets. There are dog experiences of being abused, of being loved, of losing those the dog loves and who love the dog, of grieving, of getting over grief, of forming new relationships. And, as with human beings, there is the experience of being overwhelmed by animal instincts and of engaging in the process of learning how to regulate instincts within the overall multi-layered doggy awareness.

As regards spiritual cognition, during any incarnation dogs are never aware that they are a fragment of an ongoing spiritual identity.

This marks a difference between the quality of awareness of spiritual fragments that occupy human bodies and those that associate with canine bodies. This, we must point out, does not mean that one type of fragment is superior to another. It just demarcates a difference in fundamental nature.

Of course, there are other kinds of spiritual identity who maintain a much greater awareness of their spiritual identity while embodied than those who incarnate in human bodies do. In this sense, they are more complex fragments than you are and we were. We note, there are none such incarnating on this planet. They occupy bodies in other parts of the multiverse.

We state this merely to repeat our point: there is no basis for human beings to get big-headed or filled with ideas of superiority. A range of complexity exists among nodes and the spiritual identities they form. Human beings are neither the simplest or the most complex form of identity. Human beings are merely one variety among an extensive range. So the conditioned human sense of superiority to other species on this planet, and the sense that the human is closer to God than any other identity, is simply wrong. We advise all to readjust their outlook to accommodate this fact. Pet lovers, of course, already know this through their personal experience, even though they are unlikely to describe what they know in quite these terms.

CHAPTER 11

LIFE IN THE MULTIVERSE

THE WORLD VIEW WE ARE PROMULGATING here is an expansive, multiverse view. The human view is narrow and parochial. The universe is incredibly diverse, containing life forms of every imaginable kind. Some life forms are made of minerals. A human being looking at them would not consider them to be alive. Other life forms are wispy and, again from a human perspective, are too nebulous and dispersed to be viewed as a single living being. Many of the weird and wonderful beings science fiction writers concoct have a correlate somewhere in the multiverse.

You, our reader, might assume that we are asserting this knowing there is no way you can verify what we are saying. In fact, the opposite is true. We say it because we know many people have in the past, and are currently today, verifying the validity of this proposition via personal perceptions. Accordingly, we offer this information precisely because we know it can be verified via empirical observation.

For many people, discussions of other planetary life-forms leads to the question of whether extraterrestrial-biological entities (EBEs) are zipping around the Earth's skies in spaceships. We affirm this has occurred, but not to the extent and in the ways that many think.

A fundamental limitation when discussing such matters is that human beings project preconscious-level assumptions onto the universe, and those assumptions are shaped by their attachment to the human bodily form, which is their default reference point in any discussions of this topic. So just as human beings spend their days driving around in tin cans they call cars, so they project the notion that EBEs are flying between worlds in tin cans called spaceships. This betrays far too little imagination. The EBE experience is much more diverse and interesting than this.

Extraterrestrial biological entities exist within the multiverse on all the frequencies we have differentiated into the electrophysical, electromagnetic and electrospiritual. To understand the difference between a human body and, by way of an example, one kind of EBE body, consider that fog contains physical molecules just like your body, but they are much less densely packed than in the human body and are suspended in air. Similarly, this EBE body is much more subtle than human bodies, consisting of physical molecules that vibrate at a high frequency and possess so little physical density that their bodies may be likened to gas molecules suspended in ultraviolet light. So where the human body predominantly manifests as electrophysical, with electromagnetic impulses keeping it functioning and with its electrospiritual functions not consciously connected to its everyday identity, the EBE body we are identifying here is predominantly electromagnetic, contains wisps of the electrophysical, and its identity consciously utilises its electrospiritual functions.

This EBE body doesn't need to ingest food and drink like the human body does to continue living, so it doesn't have respiratory, digestive or excretory systems. The body shape is also less defined. Human observers have noted the degree to which some EBEs are able to shape-shift. This is because they project a mental image of themselves to those human beings to whom they choose to appear. The less an individual's awareness is tied to the outright physical, the more activities are actioned by pure intent, and the more the individual is released from a single physical identity.

Throughout the physical multiverse EBEs exist in a wide range of forms and frequencies, from the incredibly dense, being rock-hard and apparently immobile, to the biological, as on this planet, to what may best be termed ethereal. From the human perspective, being a biological entity involves carbon-based life-forms fuelled by water and light. From our perspective, a biological entity encompasses every type of life-form imaginable, evolved from any and all chemical bases, and living within often inconceivable environments. An example of what we mean exists here on Earth, in the creatures that scientists have discovered living in boiling water heated by undersea volcanic vents. Scientists considered life in such environments impossible—until living creatures were found. The same may be said for life in the multiverse. It exists in forms and at frequencies within the electrophysical and electromagnetic spectrums that human beings cannot currently conceive. Of course, just because they are inconceivable doesn't mean they do not exist out there. Or are not already here.

We offer this last statement with our tongue metaphorically in our cheek, because you, as a spiritual identity, are an extra-terrestrial. You come from elsewhere. The same applies to every spiritual identity on this planet. None originated here. All originated in another dimension, or, more exactly, on another bandwidth, given reality is one, it just exists as a continuous, overlapping series of bandwidths.

Accordingly, all spiritual identities on this planet are, as is said in human terms, a migrant. Reader, this is not your home. This planet is a place to which you have travelled for the purpose of experiencing, learning, growing and developing. This exact same purpose is behind all the other spiritual identities present on this planet. They are here to experience, learn, grow and develop, each in their own way. However, this is not the only physical place to which you may migrate.

Every spiritual identity now incarnated in a human body has seen other worlds and enjoyed other non-terrestrial states. This is one option available between lives: to explore other places. It is even possible to physically associate with the creatures living there. As a result, you may have associated with creatures on other planets. It is not certain that is the case for each of our readers, but it is a significant possibility.

Why would you travel to another world, whether for the equivalent of sightseeing, or to associate with creatures non-human in nature? For some it is curiosity, to see something different. After all, incarnating life after life into the human world is demanding. Going somewhere else to have different, less intense experiences can be refreshing, cleansing. So that is one reason individuals choose to visit other worlds.

Another is to test particular faculties, for example, telepathy, or artistic ability, or empathy. Cultures in other worlds have different social structures to the human, and different balances of cognitive functions. So going elsewhere can provide an opportunity to develop an ability in a social environment that is less demanding and more supportive than is offered in the human world. This is a second reason you may have travelled elsewhere, to undergo what, in effect, is a short-term workshop to enhance your skills.

These kinds of travel are not random. Just as there are migratory routes across this planet, along which people travel from one place to another before arriving at their final destination, so there are migratory routes through electrophysical and electrospiritual domains. Experienced node identities who share your basic spiritual nature have explored alternative places and so are in a position to make recommendations to you based on what they know of those places and what they know of you. Their knowledge and expertise enables them to function as guides, helping the less experienced understand what is beneficial to them and pointing out the virtues of the available possibilities.

For an extended visit to an alternative physical world to be possible, a crucial prerequisite is that the cognitive and cultural configurations of the creatures on that other planet resonate with you. This is because you, as an individual incarnating in the human world, have accumulated particular human-centred experiences and skills. So there has to be a sound fit between your expertise and the new world you are visiting. If you travel somewhere where totally different knowledge and skills are utilised to those you have accumulated here on Earth, you will struggle to connect your identity with the body with which you intend to associate. Neither will you be able to interact effectively with the physical and social environment it inhabits. So unless you wish to make a serious, multi-life commitment to learning how to exist in a world totally different from the human—a commitment some do actually make, but so rarely as to be negligible in terms of what we are discussing here—the world you select is required to resonate substantially with your accumulated human cognitive capacities and with the skill sets you have developed.

This means it is only when you reach a certain level of expertise that you possess the capacity to take advantage of what is available elsewhere. This applies to physical as well as non-physical worlds. In order to travel on this planet you need a ticket; without a ticket you don't get through the door. Similarly, to travel to other worlds you need a ticket, which is your accumulated developmental state. On Earth you work at a job to earn money to buy a ticket. On the level we are referring to here, you work on your identity to earn entry.

What we are attempting to show is that you, as a spiritual identity, have one main base, this planet, where you incarnate regularly in order to experience, learn, grow and evolve. However, you also occasionally travel to other worlds. Some you visit for time out and to relax, and some you visit to engage in inner work. The first treats other worlds as holiday places, the second as places you travel to to attend workshops and learn something new.

Naturally, we are aware that this may be read as science fiction. We can't deny this. But why do you think science fiction is so popular? Note, however, we have not mentioned flying saucers. When, between lives, you go to sightsee in another world, you do so entirely spiritually, without a body. So you have no need of a physical vehicle. Similarly, when you attend a workshop in another world, this may involve associating with a body, but there are several ways of attending. You may incarnate as an individual in a body, but this is uncommon. What is more usually the case is that you will piggy-back on another incarnated being, who gives you access to their awareness as they experience a situation on that planet via their own. Or you will be present electrospiritually, with impressions arriving via your energetic body rather than through your physical senses. Often individuals do the equivalent of auditing, being present but remaining an observer only, and not taking part directly in what is happening.

What, then, about other spiritual identities? Just as human identities go to other worlds to sightsee, to attend workshops, or to experience physical life there, do other spiritual identities, for whom the Earth is not their main base, come here to do the same? As would be expected, the answer is yes. Just as other worlds are on migratory routes for human identities, so this planet is on other identities' migratory routes. But, to reiterate, no one turns up by chance. Just as you have a reason for going to other planets, so these identities have a reason for being here. As you would also expect, their reasons are various.

This planet offers opportunities for experience and learning. Spiritual identities domiciled in other worlds use visitation to this planet to further their own growth. However, not just any identity can do so. The prerequisite is compatibility of experience, cognitive abilities and skill sets. So just as you can only travel to and participate in worlds that offer a fundamental compatibility, so other identities coming here to interact and participate have to be compatible cognitively and socially, and need to possess a similar level of skills. There is no utility in an identity who has the musical development of a maestro incarnating on a planet where everyone is at the level of playing *Chopsticks*. Or visa versa.

That non-human spiritual identities are here occupying human bodies means several things. First they are not really non-human. The two sets of identities have much more in common than not. They are sufficiently similar to be what is termed kissing cousins. Second, that they have found their way here means that exploring nodes have scouted out the possibilities and made the match. As a result, this planet is on the migratory route of these non-human individuals. Third, this means they haven't started coming here just recently. They have been doing so for millennia, just as individuals who identify as human have been visiting other worlds for millennia.

However strange all this sounds—and we are aware it will sound exceedingly strange to those for whom these ideas are new—that is what is happening. Visiting planets that are not a spiritual identity's home base is an entirely normal activity for those who wish to test themselves, develop their current skill levels, and evolve as fragments or nodes of Dao-identities.

The reason we have broached this topic is because we wish to indicate how life on this planet is connected to a much more expansive range of activity. Human existence is connected to, and embedded in, the interpenetrating electrophysical, electromagnetic and electrospiritual levels of the multiverse. What happens on Earth impacts on others existing elsewhere.

So it is inadequate to view life on this planet, as scientists do, as a one-off, accidental occurrence that will soon subside into oblivion. It is also unsatisfactory to consider, as the religious do, that what happens on Earth occurs at the whim of a God who engineered it long ago and who continues to control everything today. Life on this planet exists because nodes of Dao-consciousness nurtured it into existence, and other nodes now occupy it for their education. Furthermore, everything is overseen by identities who are motivated by agapé, and so are deeply concerned whenever something happens on Earth that is detrimental to swathes of its species populations.

In this context of concern, we can say that the single over-riding quality that is currently lacking among humanity is a sense of responsibility. Certainly, many individuals are acting responsibly within their spheres of influence. But much of the humanity's planetary leadership is behaving like tetchy, self-obsessed teens. And their followers are behaving like capricious privileged children who think the worst response is they will be sent to sit for a while on the naughty chair. They are failing to take their life sufficiently seriously.

That is why we are now appealing to those who, as is colloquially said, are the adults in the room. To you we offer this mental picture of humanity as having a place within the multiverse, functioning on multiple bandwidths of reality, in order to bolster your sense of deep connectedness and consequently your sense of responsibility.

We encourage you to adopt a wide context when pondering your existence. We also offer assurance that you have support in your efforts to apply yourself and grow. And we emphasise that your efforts are not just for your own sake, but for the sake of every other identity currently existing on this planet, and for the manifold identities who see this as a home away from home, whose connection with this planet will continue far into the foreseeable future. And beyond.

AFTERWORD

THE ARCHIVING OF THE MATERIAL RECEIVED TO DATE has created an asset which is worthy of others' study. Its editing into this book provides a straightforward introduction to difficult concepts, many of which are outside the ordinary person's experience. Yet this material has the potential to introduce people to a modern conception, and therefore a more accurate interpretation, of their experience, if and when they choose to direct their attention towards their spiritual self, which is their fundamental being.

To this end we have brought our two collaborators into association, and stimulated their continued productivity, as part of our intent to assist those who have forgotten their true nature. We hope this brief text goes some way towards helping them achieve their aims.

FURTHER READING

In the following, Keith Hill outlines what the other channelled books he and Peter have produced deal with. The books have been written to present the guides' teachings from introductory to advanced levels. The books provide three kinds of material: metaphysical, developmental and discussions of psychospiritual skills. Reincarnation is axiomatic to the guides' teaching and underpins each book, which collectively may be said to explore the implications of reincarnation from both embodied and non-embodied perspectives.

All books are available for purchase at www.attarbooks.com.

THE CHANNELLED Q+A SERIES

This series presents easy-to-read, non-technical books that explore spirituality at an introductory level. For each, the guides set a general topic for discussion, then had Keith, this series' channeller, invite people he knew to submit questions, which the guides then answered. Each book contains twenty-one questions and answers, with a number of questions picking up and expanding on prior answers. The result is a spontaneous to-and-fro in which, as the guides comment, many surprising and unanticipated topics, from their perspective, are explored. On the other hand, the guides also surprise with their answers. As of this publication there are three books in this series. While they are listed in the sequence they were created, they are designed to be read as stand-alone texts, and in any order.

WHAT IS REALLY GOING ON?

The first book in the CHANNELLED Q+A SERIES begins by considering the implications of reincarnation on human existence. It then explores a range of fascinating topics, including how human life is structured, whether pets have souls, the nature of God, if God punishes us when we screw up, whether gut feelings offer genuine spiritual insights, where evil fits in, if there is a spiritual hierarchy, is more than one of us here, if reincarnation really occurs what happens to us between lives, and are we just a cosmic joke?

The answers come in direct, non-technical language, and often have a tongue-in-cheek sense of humour. Here's an excerpt that is typical of the approach adopted throughout this series.

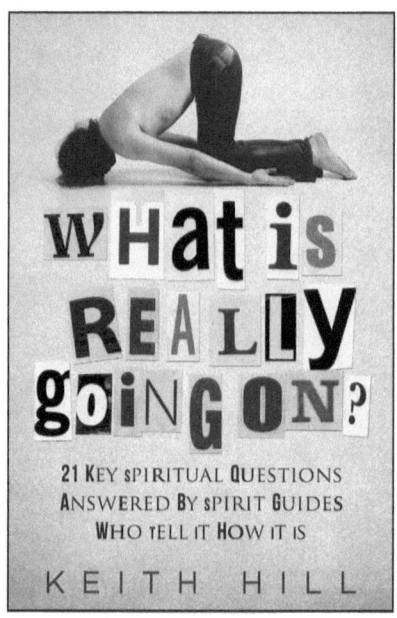

"What we undertake to do in this book is to clarify the most important facets that make possible what you experience as 'the ride of your life'. In relation to this Paul of Tarsus stated: *When I was a child, I spoke as a child, I understood as a child, I thought as a child; but when I became a man I put away childish things.* An analogy will indicate why you don't see clearly.

Imagine you're a child in a fun fair. There's such exciting stuff to do: the roller coaster to ride, the ghost train where you can scream in the dark, the coconut shy to throw balls at, hot dogs to wolf down. And of course you're not there by yourself. You're with your family, maybe a bunch of friends. A huge part of the pleasure of being at the fun fair is sharing the experience with those you care about.

On the other hand, many factors make your presence at the fun fair possible, factors that you're not aware of at all. If it's a travelling fun fair, a lot of organising has occurred even before it arrives in town. Schedules are locked down months, even years, beforehand. Workers are signed up to transport equipment and to set up the tents, machinery, fencing, ticketing offices, and signs. New rides are commissioned, old rides have maintenance done on them. People are trained to perform the myriad tasks that keep the fun fair operating on time and in accordance with safety regulations.

The same happens with life. There is a lot of behind-the-scenes organising. Those living their life are in the same situation as children at the fun fair. They don't see what it has taken to get their experiences up and running. They don't comprehend what it has taken to initiate the ride that is their life. In what follows we aim to clarify and illuminate, little by little, so your understanding may grow from childish to adult. And the first notion we need to introduce in order to dispel a fundamental layer of confusion that definitely darkens the glass through which you perceive your life is that of reincarnation. Appreciating the role reincarnation plays in your life is basic to understanding what is going on."

WHERE DO I GO WHEN I MEDITATE?

This second book focuses on key aspects of meditation. It doesn't so much deal with how to meditate (numerous books already do that), as with how meditation may be used to learn more about your purpose and to obtain information crucial to achieving fulfilment.

Topics include using meditation to find God, why people see figures such as Jesus or Buddha, how to tune into subtle communications, the role of chakras in meditation, who we typically encounter when exploring spiritual realms, whether or not protection is required, the significance of intention. Surprise topics include a consideration of extra-terrestrials and if all of us is here.

What follows is excerpted from an answer regarding what is to be gained when people use meditation to learn more about themselves.

"We have previously observed that you are only given information about yourself on a need-to-know basis. Only as information becomes

significant to you—because you need it to solve a problem, illuminate a difficulty, or require a broader context to understand what is happening—is that information released. It isn't that secrets are arbitrarily withheld, but simply that key information only becomes important to you when you need it to advance. If you received the information too early it would be confusing. Receiving it when you need it ensures it strikes home vividly and tellingly.

The same principle applies to what we are discussing here. Embedded in the call to journey into the spiritual realm is a question, a lack, or a problem to be solved that is germane to your current life situation. So when you respond to the call, gather your inner resources, step over the threshold, and project your awareness into the spiritual domain, you are doing so because you are driven by the need to discover something necessary for your continued growth.

One of the most commonly felt reasons for entering the spiritual domain is that explorers wish to encounter the Divine. Whether the Divine is envisaged as a personal being, a particular god-form, a formless field of energy, or a limitless intangible spiritual presence, throughout history seekers have sought a spiritual encounter with what is greater. The motive is usually a passionate desire to transcend the limitations, conflicts and dissatisfactions basic to human existence. In fact, what they are seeking is an encounter with their own spiritual self. This is the first level of spiritual interaction. And it is a very rich encounter. We'll explain.

Imagine that you had access to an individual who remembers all they went through during hundreds of lifetimes, has collated all their experiences into an archive of knowledge, and has extracted innumerable life lessons from everything they had undergone. Moreover, this individual is not caught up in the hurley burly of daily life, but is able to maintain a level of detachment while also being compassionate, understanding and tolerant. Such an individual would certainly be considered wise in human terms. Among the spiritually inclined, such a person is called a guru, an inspired and inspiring teacher.

You can certainly go on a physical journey to meet such a person. However, what you must know is that a wise person is available to you without physically travelling anywhere. That wise person is your own spiritual self. The inspired and inspiring guru is you. You have lived hundreds of lives. You have gone through multiple experiences. You have a huge stock of life lessons to be drawn from. And you don't have to cross oceans or mountains or wait among thousands of other seekers in order to have a personal audience. You are already there. All you need do is sit quietly, close your eyes, cross your inner threshold, and encounter your own expansive, transcendent self."

HOW DID I END UP HERE?

The third book explores life plans and how we end up as the person we are, experiencing the situations that dominate our existence this time round. Our life plan incorporates what body and family we are born into, what major challenges we face, the key people we will meet who will help us achieve our goals, and agreeing to meet others in order to help them fulfil their own life plans. The book also explores using self-enquiry to understand our life plan and ourselves.

Topics include: Why appreciating the key aspects of your life is helpful; why God isn't driving our existence; the use of psychological analysis to illuminate both our life plan; and what stops us from fulfilling our life plan. Other topics: Why gurus say different things; psychedelics; diet and phobias; the nature of other life in the universe; and why life often feels so hard.

The following is an excerpt from the answer to a question regarding can investigating our psychological make-up really be useful for understanding our life plan.

"In industrial circles researchers regularly use a process called reverse engineering. This occurs when one country downs another country's more advanced military aircraft, or when a competitor steals another company's prototype. Technicians then examine the piece of hardware and analyse it to discern not just how it was physically manufactured, but to understand the combinations of scientific principles that were utilised to construct it and make it operative.

This is the same process we are advocating here: that you, in effect, reverse engineer your current life to understand the deeper factors that led to your current talents, weaknesses, behaviours and character traits being present within you. You do so by examining the circumstances of your life and analysing them to discern what led to you being the person you are, living in the circumstances you are, making the decisions you are.

A question could be asked here. Most spiritual seekers have been told that by lifting the veil, by looking beyond the physical and social aspects of their existence, they perceive the deeper spiritual reality that underpins all existence. Yet we are saying something else. We are saying that what you discern are psychological traits, and these are certainly already present in everyday life. So where is the spiritual mystery we promised earlier? Where is the deep penetrating insight into the truly profound aspects of existence?

This is an important question. We raise it because answering it provides another significant insight into the nature of your life. First, you are at all times a spiritual identity. At any moment you can lift your awareness out of the circumstances of your daily existence to perceive reality in entirely spiritual terms. At least, that is a possibility. Yet it is a possibility not many people take up. Indeed, few people know they can step beyond their everyday human self and embrace their non-embodied spiritual identity. Many deny such an identity exists, while others believe they can only engage with it after their body dies. And even among those who accept they are a non-embodied identity, few possess the skills to regularly engage with it. Why not? What is the difficulty? The difficulty is that most people are intensely engaged in what they incarnated to do: they are engrossed in the complexities of their everyday existence.

Human existence is demanding. All the people you interact with, all the circumstances you face—from simple to tricky, from mundane to profound—require your close attention and so grab most of your daily energy. The demanding nature of human existence makes it very difficult for you to shift your attention to the veil, let alone look beyond the veil and perceive the deeper reality that underpins your physical existence. What is required is that you gather your energy and use it to energise your attention so it may be lifted out of a state of being constantly caught up in everyday existence. And you can only do that by learning what is holding your attention within everyday existence. In other words, before you can engage with deep spiritual reality you need to study and understand all the factors that "glue" your awareness to life circumstances.

Accordingly, in order to develop spiritual insight you need to dismantle what is tying you down and preventing you from attending to deeper matters. This is why we recommend self-enquiry. It enables you to unshackle your awareness so you are no longer wholly caught up in the minutiae of human existence. It helps you to broaden your perspective and deepen your insights."

METAPHYSICS FOR THE TWENTY-FIRST CENTURY

Wherever we go we need maps. Maps give us a sense of where different places are, and how we can travel from one to another. Metaphysical descriptions serve the same purpose. They map the unknown terrain that surrounds us, they illuminate the hidden forces and structures that shape the world in which we live, and they indicate the nature of alternative spiritual realities. Historically, religions have supplied humanity's metaphysical descriptions of reality. But today these religiously designed maps have become outmoded. The sciences have uncovered much about the universe, this planet, and the nature of being human, of which our forebears had no concept. Accordingly, we need new metaphysical descriptions of reality that chime with current scientific knowledge. The guides offer precisely this in the following books.

THE KOSMIC WEB

This book offers fuller explanations of many concepts presented in introductory form in this book. It begins at the beginning, with the creation of the multiverse, then discusses the seeding of ecosystems, biological life and humanity. It goes on to explore many of the concepts introduced in these pages, but in much greater detail, specifically: the nature of Dao, the nature of the electrospiritual, the function of the aura, and the evolution of nodes and node fragments. It concludes with a consideration of the traditional Great Chain of Being, which is updated into the concept of the web of life. *The Kosmic Web* is recommended as the primary text for appreciating the guides' view of existence.

"The nature of the body, including its basic shape and functioning, is well known. What is not well known is the nature and shape of the spiritual identity that locates itself beside, within and through the body. The individual spiritual identity, and we are talking now in an energetic sense, exists in the shape of a sphere. Its structure is globular. Imagine a ball, say a soccer ball, with the hexagons and pentagons forming a net around its surface. Now imagine every intersection on that ball connected radially both inwards and outwards, forming a three-dimensional structure consisting of cellular interconnecting filaments. Here the image of the hexagonal walls of honeycomb inside a beehive are appropriate. The cellular structure inside the spirit sphere is patterned like the beehive, but much less rigidly, certainly not hexagonally, and the filaments extend inwards and outwards throughout the entire globular structure. The interconnecting filaments provide the means by which encoded information is carried through the entire structure.

Encoded within the spiritual identity's globular structure is information regarding what has been done in previous lives and what is intended to be achieved in the new upcoming life. How is information regarding the life plan communicated by the spiritual self to its animal human self? The answer is via the aura. So when a spiritual identity despatches an aspect of itself into the human realm, the now born and growing human being can access knowledge of all the information encoded in its globular spiritual structure via the aura.

When the body dies and is incinerated or decays, the aura itself dissolves. But in the process of dissolving, all the information contained within the mind of the individual's human physical self is spontaneously uploaded, to use that modern concept, as patterns of information conveyed to the globular spiritual identity through the connecting link of the aura. This is easily achieved, because the identity is bi-located, with its spiritual self and its bodily self nested beside and inside one another. There is no distance for the information to travel. The uploading is direct, spontaneous and instantaneous."

THE MATAPAUA CONVERSATIONS

This book is the first collaboration between Peter and Keith. The circumstances that led to its creation is that after editing Peter's *Guided Healing*, Keith saw an opportunity to exploit Peter's contact with non-embodied identities to question them about the big bang, evolution, the nature of consciousness, and other metaphysical topics. He eventually came up with a list of one hundred "big questions".

Peter took time out at Matapaua Beach to receive the answers. He also kept a diary on the process. The excerpt that follows was spoken to Peter by the guides soon after he arrived at the secluded house where he stayed while channelling the answers to Keith's questions.

"We begin the developmental and systematic aspects of this teaching, fostered and communicated by the man Keith Hill. Given that it is at his request that we respond to his questions, we feel that proper acknowledgement should be made to the facilitative agreement that he has with us and with you. This agreement was made pre-life, of course.

When contemplating engagement with the earthly realm, you both felt there was an opportunity to contribute to a deeper understanding of the human condition. The opportunity was undertaken in a spirit of self-exploration and self-development. It also involved community engagement and company-inspired social development—"company" in this sense is to be interpreted as meaning a social group whose members share a curiosity concerning the dynamics of life and existence.

That said, we may continue in our task. Rereading Capra [*Editor: Peter was reading The Tao of Physics by Fritjof Capra*] is an essential first step for entering an analytical mindset and re-igniting your curiosity regarding the deep domains of description concerning both agapéic space and the realm of reality associated with the questions that you have come here to have answered.

The domain of these questions is extensive. Answering them requires that a boundary-free condition be established within your mind. That is not yet available, so we have set you the task of attending to the explanatory language offered in that text [*The Tao of Physics*] to take your awareness away from the mundane, away from the local, away from the body's senses and preferences, and expanding your awareness towards the foreign, and even the nonsensical. We use that last term in the special sense of being unrelated to the senses.

The mind is, in fact, free to soar into any realm at any moment, for any duration, within any time frame. This means that all of existence is accessible to the unconstrained mind. The frameworks of thought generated by locality, by embodiment, by emotion, by relationship—all these act to constrain the mind and its reach.

That is why self-isolation is required, to bring those unconstrained realms into view. That also gives rise to the need here to have mind-expanding literature at hand to refocus, as you did earlier when observing the night sky, because your perception is then not confined to the atmosphere. And it serves to remind that there are realms upon realms of distance, of phenomena, of locations far away, of unimaginable magnitudes, waiting the tools to make them observable. Those tools (for example, a telescope) are not available here. But the exercise was sufficient to remind him of the extent to which the world is bigger than this room, this body, these emotions, and these thoughts. The universe is correspondingly larger. The realm of all that is is unimaginably larger again. So, through this reminder, we bring that entire realm back into the central awareness as the vista to be explored during our time here."

AGAPÉ AND THE HIERARCHY OF LOVE

Peter's first channelled book uses a wide range of models and metaphors to explore agapé theory. The overall thrust is scientific and secular, reconceiving traditional spiritual concepts in fresh and frequently innovative ways that are consistent with our modern outlook. Topics include the nature of the higher self and its mind and how to establish a connection to it, the thousand life incarnation model, the spiritual nature of love, and an exploration of the implications of agapé theory using models of many different kinds. The following is excerpted from the book's introduction.

"We intend to initiate an international movement towards the study of the impact of the spiritual domain on human life in all its variety, and in particular its impact on the theory of being. Where once ground-level knowledge relevant to agriculture for food creation was sufficient, now theory has encompassed the origins and end of the known universe.

The question, then, is what lies beyond that? It was once enough to respond to the heart-felt love emanating from the spiritual domain, which people experienced via visions and in dreams. Now a more dynamic understanding is available that provides direct knowledge of the entities who inhabit the spiritual domain.

Once the current arguments regarding the existence of the spiritual domain are resolved, and a new consensual reality is established that draws on the empirical observations of mystics, a new inclusive outlook will arise. This will make a complete view of reality, that integrates the spiritual and the physical, available to anyone willing to take the time to develop the skills needed to perceive it. Religion in unsophisticated forms will fall away, as simplistic belief will not be required."

THE CHANNELLED SPIRITUALITY SERIES

The Channelled Spirituality Series provides for those who wish to engage in their own personal development. It provides a means for seekers to understand their individual psychospiritual make-up, what factors drive their current existence, and what the key factors of their life plan involve. In this sense, the Channelled Spirituality Series offers a practical way to carry out self-enquiry.

The series' channeller, Keith, has a background in the Fourth Way Work teaching of G.I. Gurdjieff, as developed by his New Zealand Fourth Way teacher. Because these books have been offered through Keith's mind, it builds on what he understands of the Gurdjieff Work. In addition, Keith has been directed to the Michael Teachings. These were originally channelled by a group in San Francisco, who had also been trained in the Gurdjieff Work. The Michael Teachings add many details to the Gurdjieff Work, the most fundamental being that it puts psychospiritual development into a reincarnational context. The Michael Teachings also add considerably to Gurdjieff's ideas on human psychological make-up.

Selections from all this material have been utilised by the guides in the Channelled Spirituality Series to create a straight-forward, psychologically-based approach to personal spiritual development.

EXPERIMENTAL SPIRITUALITY

This first book in the series introduces the rationale for adopting a non-religious, empirical and experimental approach to spirituality. Topics covered include how to treat beliefs as propositions and not as truths, how spirituality involves a journey from belief to understanding, and how finding answers depends on asking the right questions. The book also introduces the model of the five-layered self as an aid to understanding the human psyche.

"By now it will be clear to readers the extent to which beliefs of all kinds are naturally generated by the human everyday awareness during its daily interactions. Many beliefs are totally innocuous, such as the belief that an apple a day will keep the doctor away. Other beliefs involve practical necessities, such as the belief a light will turn on when you flick the switch. And some beliefs reflect truths about the world, such as the belief that the Earth travels around the Sun. In this last case such belief is better described as knowledge.

As was stated earlier, a particular belief is no more than an assumption until it has been tested and the tester has ascertained whether or not it reflects reality. Tested and validated beliefs become the basis of new knowledge. Because individuals have different experiences, and because people often draw quite different conclusions from the same experiences, there is an extensive range of positions from belief to knowledge. And even within this range there are those at the extremities, such as those who espouse lunatic beliefs and those who express extensively tested knowledge. This latter category applies both to knowledge obtained through objective scientific testing and to knowledge garnered as a result of interrogating one's own assumptions and outlook and tempering them in the fires of personal experience.

It is, of course, mere common sense to draw attention to this range from the lunatic to the knowing. Anyone who gives the idea a moment of attention can easily verify it from their own experience, and can further verify how the very different functioning of people's everyday awareness accounts for the huge differences between assumed beliefs and hard won knowledge. Accordingly, from the spiritual perspective advocated here,

the principal point that needs to be made is that beliefs are held by immature individuals or cultures, whereas knowledge is in the possession of the mature.

When two people argue over their belief or non-belief in God they are each indulging in childish behaviour—because they are arguing over beliefs, not knowledge. Of course, each side maintains what they believe with great intensity. Each considers they know The Truth and that those they oppose do not. But thinking you know is not the same as actually knowing. Neither is holding intensely to a belief the same as actually knowing.

Until a belief has been rigorously tested it is nothing more than an assumption. And an assumption should be considered as only a starting point on the journey towards knowledge. This is the journey from immaturity to maturity."

PRACTICAL SPIRITUALITY

The second book begins a consideration of the psychological factors that we draw in when selecting our next incarnation, along with all the other factors that are involved. Topics include: the characteristics of core disposition that makes each human being a unique spiritual identity; what is involved in selecting a life plan; seeking our bliss; the creation and resolution of karmic relationships; life lessons; and mastering the art of incarnation.

This book also extensively adds to the model of the five-layered self introduced in *Experimental Spirituality*, showing how it is an effective tool for seekers to understand their complex psyche. The following is an excerpt that discusses the contribution one's life goal makes to the life plan.

"Psychologically, every human being has a single defining essence level life goal. This life goal denotes a fundamental approach to living that the spiritual identity has adopted. It underpins everything the individual is striving to achieve during this life.

The individual draws on a number of factors when deciding how it will live its next incarnation. These include positive and negative deep essence qualities developed during previous lives, essence traits innate to the next body's genetic disposition, karma that is to be worked through, and agreements to meet certain other incarnated people to help achieve mutually agreed tasks. All these are part of an individual's life plan.

Another key contributor to the life plan is the individual's life goal. The life goal is a psychological characteristic that functions at the essence level. There are seven in total. We note that the life goal is not the same as what is generally referred to as life goals.

Life goals are specific pragmatic physical and social goals individuals set themselves to achieve during the course of a week, a month, a year, or even a lifetime. Life goals include taking part in a competition, establishing a career, having a family, becoming wealthy, dedicating one's life to helping the underprivileged, becoming a country's president or prime minister, or not being like one's parents or siblings. Life goals may be large or small. But they are physically or socially defined, and they all involve a series of steps or tasks that need to be accomplished in order to ensure the chosen life goals is achieved. We call these life tasks.

In contrast, the *life goal* doesn't involve a particular task or series of tasks. The life goal may be thought of as psychological glue that underpins all the various aims, tasks, agreements, karmic debts, and essence activities that the individual chooses to carry out during the course of a life. As such, the life goal is aligned to each individual's life plan. The life goal defines the individual's overall approach to a life. Drawing on the Michael Teachings, we identify seven fundamental types of life goal. These seven are: growth, revaluation, dominance, submission, acceptance, rejection and maintaining equilibrium.

Each of these identifies what, in global terms, an individual seeks to achieve in a particular life. It identifies an individual's approach, that is, an individual's essence level psychological predisposition, when faced with choices and alternative possibilities during any life."

PSYCHOLOGICAL SPIRITUALITY

This book is paired with *Practical Spirituality*, as the guides required two books to fully explore the impact of reincarnation on human psychology. The concept of accumulated human identity is introduced to account for what is uploaded from a life to the spiritual self, then used in

a later life. Other key concepts include: the role of fear in shaping childhood behavioural coping mechanisms that extend into adulthood; how fear lays the foundations for chief feature, the conflict between true and false personality; how the nature of life plans progressively change through our entire cycle of incarnations; how to change our psychological momentum; the need to think of ourselves as multi-life identities; and the key factors of orientation and attitude.

"To begin this discussion of how false personality stops you from fully achieving your life plan, let's summarise your life situation. You are born into a body that has certain genetic predispositions. The family and social environment in which you are raised imposes specific attitudes, beliefs and behaviours on you. Because human existence is complex and difficult, you develop defensive behaviours to help you cope with trying situations. These self-defensive behaviours are gathered around a chief feature that has a specific fear at its core. Over time these fear-based self-defensive traits become imbedded in your socialised self.

You also have an essence self. Your essence self is an expression of your spiritual self. It is the higher human part of you that is seeking to fulfil specific tasks, develop skills, foster abilities, and express talents. It is the best of your human sub-personality this time round. But in order for you to work with other people, your essence self has to express itself via your socialised self.

Your socialised self possesses a number of neutral traits that enable you to interact effectively with others: language, customs, socially agreed ways of behaving. But it also has a dark side, consisting of your deep fears, your chief feature, and buried defensive attitudes. Collectively, these dark aspects of your socialised self are inhibitory mechanisms that hinder your essence self's efforts. We are calling these fear-centered, self-defensive, inhibitory traits false personality. Psychologically, they have their own momentum within you, their own inertial mass. And this momentum leaks into and impacts on your essence self.

As a result no one's life momentum is optimal. No one is implementing every single aspect of their life plan to its maximum potential. We don't say this to deflate you. Far from it. We say it because the fact is that no one lives their life without at least some negative impact from false personality. Everyone has experienced times when fear, or lack of confidence, or over-confidence, or second-guesssing, or defensiveness, or any of many other traits, caused them to hesitate and not do what deep down they wanted to do. This is how false personality causes you to depart from your life plan—not by offering you an equally compelling new goal, but by stopping you making progress at all."

OTHER RELATED BOOKS

The following are largely channelled books that explore various specialist areas of the materials covered in the other books.

GUIDED HEALING

This book contains two urgent messages. One offers guidance to all spiritual seekers, the other is addressed specifically to those who wish to learn about spiritual healing.

To spiritual seekers, *Guided Healing* presents a novel view of the spiritual purpose and benefits of being born into a physical body. Issues covered include the relationship between the spiritual and physical realms, the reason for incarnation, the use of meditation as a means for exploring the spiritual realm, the pluses and minuses of doing so, and the significance of soul work.

To healers, *Guided Healing* offers instruction on how to become a conduit for healing energy that emanates from the spiritual realm. Topics covered include how to contact guides in the spiritual realm, the nature of spiritual perception, and factors which enhance or hinder energy flow during the act of healing.

"Guidance enables mental images of actions to be made manifest via the structure of the healer's body. In considering this, there are two possibilities. Either guidance is via images projected into the healer's mind, or it is through direct control of the body—and here we speak of those actions performed by the healer's body, of bringing the hands into proximity with the patient's energetic structure, so remedial healing may be carried out.

If the healer has sufficient trust and is an embodied human, then the healing may be directed via the mind as a sequence of images given to the healer to emulate, by adopting the positions shown in the provided images."

PEOPLE OF THE EARTH

When Peter Calvert gathered a small group of meditators and set them the task of opening their minds to whoever arrived, he didn't anticipate the astonishing encounters that would result. *People of the Earth* offers a unique account of communications with spiritual identities normally invisible to us. They discovered a small percentage of people become confused after death. Lost in a transitional zone, they need guidance to move on to the next phase of their existence. They also communicated with nature spirits. In a series of vexed visits, nature spirits pleaded with the meditators to share an urgent message with humanity regarding the dire state of the planet's ecosystems. Finally, they encountered a number of non-

human beings who "dropped in" to see what was going on.

Through a sequence of dialogues between the meditators and visiting beings, *People of the Earth* provides sobering insights to those who seek to understand what is required of us spiritually to sustain the planet's ecological health. Throughout the text, Peter's guides add comments on those they are meeting, situating what they encounter into a wider context. The following is from the start of an encounter. Channelled material is in *italics*.

P Welcome. What have you to share with us?
I'm just looking.
Why?
You seem different.
In what way? Using what for comparison?
I was just drawn to you.
Please share something about yourself. What is your nature?
I'm a level above you.
What does that mean?
I'm not human any more.
So you were?
Yes.
Did you live more than once?
Yes.
How many times did you live?
Four hundred and fifty-seven.
Why is it that you are no longer human?
I moved on.

INDEX

Accumulated human identity 40, 41, 43, 47, 48, 50, 53, 64
Adult (development) 34, 42, 51-52
Agapé 78-81, 84-86, 96
Agapé frequency 78-84
Agapeic space 70, 72, 74, 78
   Three axes 78-81
Agreements (life plan) 37
Anomalies 59-61
Antagonist 46
Attachment 20, 48, 90
Aura 8, 30, 31, 43, 61-69, 106
   Two levels of information 63

Basement (psychology) 45, 46, 53, 60
Beliefs (limitation of) 13, 16, 59
Big bang 17, 20
Biological life 23
Biological self 41, 42, 67
Bohm, David 21, 62

Canines 87-89
Capra, Fritjof 108
Cetaceans 26, 86
Childhood 44, 45, 53, 72, 76
Choice 29, 37, 40-44, 51, 53, 57, 75, 76
Co-association (of spirit with body) 28-30, 86
Consciousness 17, 22, 24, 25, 27, 28, 29, 30, 32, 33, 38, 39, 43, 44, 47, 48, 57, 64, 78, 80, 81, 84, 86, 87, 96
Coping mechanisms 44-45
Creation 17, 20, 87
   Fluctuation 21
Cues (spiritual) 43, 54

Dao (the) 17-24
Dao-consciousness 25-32 (also see Nodes)
Data 12, 22, 58, 59, 60, 61, 64
Deist 20
Development 25, 27, 28, 34, 35, 40, 42, 44, 63, 64, 78, 79, 80, 95
Divine 13, 16, 64
DNA 22, 42
Doggy identity 88
Dream 60, 61, 66
Dreams 54, 60, 61, 81

Electromagnetic 21, 43, 91, 92, 96
Electrophysical 20, 21, 22, 91, 92, 93, 96
Electrospiritual 20, 21, 22, 43, 62, 63, 91, 93, 96
Embryo 30, 63
Emergence 17, 20, 25, 62
Empirical approach 12, 59, 77, 90 (see Data)
Energetic self 41, 42, 43, 67
Environmental niches 20, 23, 27, 29, 84, 85

Essence self 41-43, 46, 57
Everyday awareness 43, 49, 50, 52, 67
Everyday mind 54, 60, 66, 69, 72, 73, 74, 76
Experiment (existence as) 16, 19, 20, 22, 23, 24, 29-35, 38, 39
EBEs (extra-biological entities) 90

Family 26, 42, 44, 55, 86
Fear 29, 44, 45, 53, 57, 61, 65, 72, 73, 74, 82
Five-layered self 41, 52, 58, 111
Foetus 21, 22, 30, 62, 63, 64
Forgetfulness 48
Frustration 49, 56

Globular spiritual structure 30
God 17-20, 70, 77, 87, 89, 96

Hara dantian 30, 80
Hierarchy 78-84, 86
Higher mind 74, 75
Holons (Wilber) 22
Homo sapiens sapiens 28, 41, 87
Horses 26

Identification 48, 84
Implicate order 21, 22, 62
Infant (development) 44, 50
Information 21, 22, 23, 24, 26, 27, 30, 35, 41, 43, 52, 53, 54, 56, 59, 60, 62, 63, 64, 65, 66, 67, 68, 69, 74, 76, 90
Journal (self-enquiry) 54, 58
Jung, Carl 45

Karma 15, 28, 41, 46, 53
Kernel (spirit sphere) 30-31

Life lessons 23, 34, 35, 36, 37, 41, 44
Life plan 35-38, 41-47, 51, 52-57, 61, 66
Love 17, 27, 34, 35, 38, 41, 57, 68, 75, 78, 79, 87, 88

Maturity 28, 31, 50, 86
Meditation 14, 15, 54, 70 72, 76-81
Migratory routes 93, 95
Multipersonhood model 53
Multiverse 18-24, 29, 32, 84, 89-92, 96
Nodes (of Dao-consciousness) 22, 25-44, 47, 48, 49, 52, 54, 56, 57, 63, 64, 71, 78, 80, 81, 84-89, 93, 95, 96
  cast from the Dao 25, 26, 84, 86
  enrichment of 31, 33
  naive nodes 34, 35, 86
  occupancy of human species 28
  reintegrated nodes 15, 80, 86
  uploading data 43-44

Observer 17-24
Obstacles 37, 38, 45, 46, 47

Preconscious processing 76, 82
Psychic abilities 68
Psychic anomalies 61
Psychological traits 15, 22, 35, 36, 39, 40-47, 54, 57, 58, 59, 63, 64, 79, 87
Psychology (basement) 45, 46, 53, 60

Reincarnation 15, 33, 34, 36, 44, 45, 89, 93, 95
Reintegrated nodes 15, 80, 86
Responsibility 27, 31, 51-52, 72, 96

Science fiction 32, 90, 94
Scientific naturalism 19
Self-calming 60
Self-defensive 60
Self-defensiveness 60
Self-enquiry 54-58, 60, 64
Shadow self (Jung) 45
Shamanic flight 70-73
Shamanic space 70-73, 80, 82
Socialised self 42, 44, 45, 57, 88
Solo self-study 58
Spirit sphere 30, 43
Spiritual domain 11, 28, 29, 43, 56, 67, 69, 78, 110
Spiritual self 31, 41, 42, 43, 49, 52, 53, 54, 56, 57, 60, 61, 66, 72, 73, 74, 97

Teen (development) 34, 45, 50, 51
Traits (negative) 36, 38, 45, 46, 47, 52, 53, 59

Universe 17-24, 49, 87

Wilber, Ken 22
Willingness to bequest agapé 79, 80, 86
Working in a group 58
Workshops 52, 94, 95

www.ingramcontent.com/pod-product-compliance
Lightning Source LLC
Chambersburg PA
CBHW030451010526
44118CB00011B/881